ADVANCES IN SATELLITE METEOROLOGY

ADVANCES IN SATELLITE METEOROLOGY

A.I. Burtsev, P.N. Belov, and
Sh.A. Musaelyan, Editors

Translated from Russian by A. Baruch and D. Lederman
Translation edited by P. Greenberg

A HALSTED PRESS BOOK

JOHN WILEY & SONS
New York · Toronto

ISRAEL PROGRAM FOR SCIENTIFIC TRANSLATIONS

Jerusalem · London

10642

© 1973 Israel Program for Scientific Translations Ltd.

Sole distributors for the Western Hemisphere and Japan
HALSTED PRESS, a division of
JOHN WILEY & SONS, INC., NEW YORK

Distributors for the U.K., Europe, Africa and
the Middle East
JOHN WILEY & SONS, LTD., CHICHESTER

Distributed in the rest of the world by
KETER PUBLISHING HOUSE, JERUSALEM

Library of Congress Catalog Card Number 73 1981
ISBN 0 7065 1284 7 IPST, Jerusalem
ISBN 0 470 125381 1 Halsted/Wiley, N.Y.
IPST cat. nos. 22049/22050

This book is a translation from Russian of
INTERPRETATSIYA I ISPOL'ZOVANIE
SPUTNIKOVYKH DANNYKH V ANALIZE I
PROGNOZE POGODY, Nos. 73, 89
Gidrometeorologicheskoe Izdatel'stvo
Leningrad 1971

Printed in Israel

Contents

UDC 551.558.1

CELLULAR CONVECTION IN A VORTICITY FIELD

N. F. Vel'tishchev, A. A. Zhelnin

Nonlinear problems of cellular convection are treated in an atmosphere revolving with constant angular velocity. This revolution is found to impart a "stretching" action to the convective motion in the atmosphere.

Critical values of the absolute vorticity, at which hexagonal convection cells or two-dimensional billows cannot exist, are functions of the nonlinearity of the temperature profile, viscosity, and the scale of the perturbations. An increase in the nonlinearity of temperature profiles and the coefficient of turbulent viscosity, and a decrease in wavelength lead to an increase in the critical values of the absolute vorticity at which regular convection in the form of hexagonal cells and two-dimensional billows becomes possible. Quasihexagonal cells may form at an absolute vorticity of the order of 10^{-4} to $10^{-5}\,\mathrm{sec}^{-1}$. Convective cells of 100 km diameter become unstable already at an absolute vorticity of the order of $8 \cdot 10^{-6}\ \mathrm{sec}^{-1}$. This explains the absence on satellite TV photographs of quasihexagonal cells of diameter more than 100 km.

The linear theory of cellular convection in a rotating fluid has been dealt with by Chandrasekhar /7/ and Trubnikov /6/, who conclude that rotation stabilizes the process of convection. As regards the atmosphere, it has been shown that an anticyclonic situation is more favorable for the development of cellular convection than a cyclonic one, since in an anticyclone the absolute vorticity is near zero /4/.

However, the conclusions in the above papers were limited by the framework of linear theory and supplied no answer to the problem as to which form of secondary flow would be most stable in a rotating fluid.

The answer to this problem can be obtained only by examining the nonlinear aspects of convection. Krishnamurti /8/ and Vel'tishchev /3/ have shown that in a calm atmosphere convective currents are influenced decisively by the nonlinearity of mean temperature profiles appearing during the interaction of finite amplitude perturbations. The aim of this work is to extend the earlier results to a rotating atmosphere.

We assume that the fluid rotates about a vertical axis with constant angular velocity $\Omega = l + \zeta$, where l is the Coriolis parameter and ζ the vertical component of the relative vorticity. The mean vertical temperature profile is given in the form

$$\overline{T} = T_0 + T_1 + T_2, \tag{1}$$

where T_0 is the temperature at the lower limit of the layer, $T_1 = -\gamma z$, and $T_2 = \Gamma \sin k_z z$ ($\Gamma = \text{const}$).

In a simple turbulent regime this method of representing the vertical temperature change allows for heating and cooling of the convective layer as a whole, so that $\Gamma > 0$ corresponds to cooling and $\Gamma < 0$ to heating of the layer.

The viscosity (ν) and thermal conductivity ($\varkappa$) are assumed isotropic, and the Prandtl number is taken as unity ($\text{Pr} = \nu/\varkappa = 1$).

After some simplifications based on the theory of free convection, we obtain the following hydrothermodynamic equations:

$$\frac{\partial u}{\partial t} + \mathbf{V}\nabla u - \Omega v = -\frac{1}{\overline{\rho}} \frac{\partial p}{\partial x} + \nu \nabla^2 u, \tag{2}$$

$$\frac{\partial v}{\partial t} + \mathbf{V}\nabla v + \Omega u = -\frac{1}{\overline{\rho}} \frac{\partial p}{\partial y} + \nu \nabla^2 v, \tag{3}$$

$$\frac{\partial w}{\partial t} + \mathbf{V}\nabla w - g\alpha T = -\frac{1}{\overline{\rho}} \frac{\partial p}{\partial z} + \nu \nabla^2 w, \tag{4}$$

$$\nabla \mathbf{V} = 0, \tag{5}$$

$$\frac{\partial T}{\partial t} + \left(\gamma_a + \frac{\partial \overline{T}}{\partial z}\right) w - \mathbf{V}\nabla T = \varkappa \nabla^2 T, \tag{6}$$

where $\overline{\rho}$ is the standard density, $\alpha = \overline{T}^{-1}$ is the coefficient of thermal expansion, and T, u, v, w and p are functions of x, y, z and t;

$$\nabla = \frac{\partial}{\partial x} + \frac{\partial}{\partial y} + \frac{\partial}{\partial z},$$

$$\nabla^2 = \frac{\partial^2}{\partial x^2} + \frac{\partial^2}{\partial y^2} + \frac{\partial^2}{\partial z^2}.$$

This system of equations is nonlinear. In order to eliminate the pressure from (2) — (4) we operate on (2) with $-\dfrac{\partial^2}{\partial x\,\partial z}$, on (3) with $-\dfrac{\partial^2}{\partial y\,\partial z}$ and on (4) with $\nabla_1^2 = \dfrac{\partial^2}{\partial x^2} + \dfrac{\partial^2}{\partial y^2}$. Use of continuity equation (5) and simple manipulations yield

$$\left(\frac{\partial}{\partial t} - v\nabla^2\right)\nabla^2 w + \nabla_1^2(\mathbf{V}\nabla w) - (\mathbf{V}\nabla u)_{xz} - (\mathbf{V}\nabla v)_{yz} +$$

$$+ \Omega\frac{\partial}{\partial z}\left(\frac{\partial v}{\partial x} - \frac{\partial u}{\partial y}\right) - g\alpha\nabla_1^2 T = 0. \tag{7}$$

If equation (2) is operated on by $-\dfrac{\partial}{\partial y}$ and (3) by $\dfrac{\partial}{\partial x}$, the resulting equations can be combined into the form

$$\left(\frac{\partial}{\partial t} - v\nabla^2\right)\left(\frac{\partial v}{\partial x} - \frac{\partial u}{\partial y}\right) - (\mathbf{V}\nabla u)_y + (\mathbf{V}\nabla v)_x = \Omega\frac{\partial w}{\partial z}. \tag{8}$$

Application of the operator $\left(\dfrac{\partial}{\partial t} - v\nabla^2\right)$ to (7) and use of relationship (8) gives

$$\left(\frac{\partial}{\partial t} - v\nabla^2\right)^2\nabla^2 w - g\alpha\left(\frac{\partial}{\partial t} - v\nabla^2\right)\nabla_1^2 T + \Omega^2\frac{\partial^2 w}{\partial z^2} =$$

$$= \left(\frac{\partial}{\partial t} - v\nabla^2\right)\left[(\mathbf{V}\nabla u)_{xz} + (\mathbf{V}\nabla v)_{yz} - \nabla_1^2(\mathbf{V}\nabla w)\right] -$$

$$- \Omega\left[(\mathbf{V}\nabla u)_{yz} - (\mathbf{V}\nabla v)_{xz}\right]. \tag{9}$$

Operator $g\alpha\left(\dfrac{\partial}{\partial t} - v\nabla^2\right)\nabla_1^2$ applied to (6) and $\left(\dfrac{\partial}{\partial t} - \varkappa\nabla^2\right)$ to (9), and combination of the resulting two equations for $P = 1$ leads to the following expression:

$$\left[\left(\frac{\partial}{\partial t} - v\nabla^2\right)^2\nabla^2 - g\alpha\beta\nabla_1^2 + g\alpha\frac{\partial T_2}{\partial z}\nabla_1^2 + \Omega^2\frac{\partial^2}{\partial z^2}\right]w =$$

$$= \left(\frac{\partial}{\partial t} - v\nabla^2\right)\left[(\mathbf{V}\nabla u)_{xz} + (\mathbf{V}\nabla v)_{yz} - \nabla_1^2(\mathbf{V}\nabla w)\right] -$$

$$- \Omega\left[(\mathbf{V}\nabla u)_{yz} - (\mathbf{V}\nabla v)_{xz}\right] - g\alpha\nabla_1^2(\mathbf{V}\nabla T), \tag{10}$$

where $\beta = \gamma - \gamma_a$.

This equation contains four unknowns, and so another three equations are required to complete the system. The first of these is obtained by eliminating pressure and density from (2) and (3). For this purpose we apply operator $\dfrac{\partial}{\partial y}$ to (2) and $\dfrac{\partial}{\partial x}$ to (3), subtract the second equation from the first, and then apply $\dfrac{\partial}{\partial y}$ to the resulting equation. This last equation is combined with the equation of continuity and then operated on with $\dfrac{\partial^2}{\partial x\,\partial t}$ to yield

$$\left(\frac{\partial}{\partial t} - v\nabla^2\right)\left(\nabla_1^2 u + \frac{\partial^2 w}{\partial x\,\partial z}\right) + \Omega\frac{\partial^2 w}{\partial y\,\partial z} = (\mathbf{V}\nabla v)_{xy} - (\mathbf{V}\nabla u)_{yy}. \tag{11}$$

Similarly, we differentiate (2) with respect to y and (3) with respect to x, subtract the second equation from the first, and then apply the operator $\dfrac{\partial}{\partial x}$. The resulting expression is then combined with the equation of continuity to which the operator $\dfrac{\partial^2}{\partial y\,\partial t}$ has previously been applied. Hence

$$\left(\frac{\partial}{\partial t}-\nu\nabla^2\right)\left(\nabla_1^2 v+\frac{\partial^2 w}{\partial y\,\partial z}\right)-\Omega\,\frac{\partial^2 w}{\partial x\,\partial z}=(\mathbf{V}\nabla u)_{yx}-(\mathbf{V}\nabla v)_{xx}. \quad (12)$$

The fourth equation of the system is provided by the heat flux equation (6).

Before proceeding to solve these equations we shall further simplify them by taking into account the rather slow processes and disregarding terms involving time derivatives in the nonlinear parts of (10), and in the linear parts of (6), (11) and (12). In equations (11) and (12) we shall omit nonlinear terms, and rewrite the heat flux equation using relationship (1). The final system of equations now assume the form

$$\left[\left(\frac{\partial}{\partial t}-\nu\nabla^2\right)^2\nabla^2-g\alpha\beta\nabla_1^2+g\alpha\,\frac{\partial T_2}{\partial z}\,\nabla_1^2+\Omega^2\,\frac{\partial^2}{\partial z^2}\right]w=$$
$$=-\nu\nabla^2\left[(\mathbf{V}\nabla u)_{xz}+(\mathbf{V}\nabla v)_{yz}-\nabla_1^2(\mathbf{V}\nabla w)\right]-$$
$$-\Omega\left[(\mathbf{V}\nabla u)_{yz}-(\mathbf{V}\nabla v)_{xz}\right]-g\alpha\nabla_1^2(\mathbf{V}\nabla T), \quad (13)$$

$$\nu\nabla^2\nabla_1^2 u=\left(\Omega\,\frac{\partial^2}{\partial y\,\partial z}-\nu\,\frac{\partial^2}{\partial x\,\partial z}\,\nabla^2\right)w, \quad (14)$$

$$\nu\nabla^2\nabla_1^2 v=-\left(\Omega\,\frac{\partial^2}{\partial x\,\partial z}+\nu\,\frac{\partial^2}{\partial y\,\partial z}\,\nabla^2\right)w, \quad (15)$$

$$\nu\nabla^2(T+T_2)=\mathbf{V}\nabla T-\beta w+\frac{\partial T_2}{\partial z}\,w. \quad (16)$$

In order to examine the general properties of this system we effect some transformations. Using linear theory we assume functions T and w to have the form

$$\Phi(x,\,y,\,t)\cdot f(z), \quad (17)$$

where Φ satisfies the condition

$$\nabla_1^2\Phi=-k_\mathrm{r}^2\Phi, \quad (18)$$

i. e., we assume that the variables can be separated and that application of the plane Laplacian operator does not change the nature of the dependence of Φ on the horizontal coordinate. By using (14) and (15) to express u and v in terms of Φ and f, and carrying out simple transformations, equation (13) can be expressed in the form

$$\left(\frac{\partial}{\partial t} - \nu\nabla^2\right)^2 \nabla^2\Phi f - g\alpha\beta\nabla_1^2\Phi f + \frac{1}{2} g\alpha\nabla_1^2\Phi\,(f^2)' + \Omega^2\Phi f'' =$$

$$= \nu k^2 \left\{ \frac{1}{2}\nabla_1^2 \left[\frac{1}{2}\nabla_1^2(\Phi^2) - \Phi\nabla_1^2\Phi\right]\left[\left(1 - \frac{\Omega^2}{\nu^2 k^4}\right)\left(\frac{f'}{k_{\mathrm{r}}^2}\right)^2 - \right.$$

$$\left. - \frac{1}{k_{\mathrm{r}}^2} f^2\right] + \frac{1}{2}\nabla_1^2\left[\left(1 - \frac{\Omega^2}{\nu^2 k^4}\right)\frac{1}{k_{\mathrm{r}}^2} f f'' - f^2\right]\Phi^2\right\}_z -$$

$$- \frac{1}{2} g\alpha\nabla_1^2\left\{\left[\left(\frac{1}{2}\nabla_1^2(\Phi^2) - \Phi\nabla_1^2\Phi\right)\frac{1}{k_{\mathrm{r}}^2} + \Phi^2\right] f^2\right\}_z +$$

$$+ \left\{\Omega\left(1 - \frac{\Omega^2}{\nu^2 k^4}\right)(\Phi_y\Phi_{xxx} + \Phi_y\Phi_{xyy} - \Phi_x\Phi_{xxy} - \Phi_x\Phi_{yyy})\left(\frac{f'}{k_{\mathrm{r}}^2}\right)^2\right\}_z +$$

$$+ \left\{\frac{4\Omega^2}{\nu k^2}\left(\frac{f'}{k_{\mathrm{r}}^2}\right)(\Phi_{xy}^2 - \Phi_{xx}\Phi_{yy})\right\}_z . \qquad (19)$$

Here, subscripts of Φ signify differentiation with respect to the corresponding coordinate axis, while the primes of f signify differentiation with respect to z; $k = \sqrt{k_x^2 + k_y^2 + k_z^2}$ is the wave number, and k_{r} is the horizontal wave number.

Similarly, equation (16) can be transformed to the form

$$k\nabla^2\Phi f + \beta\Phi f = \Phi(f^2)' +$$

$$+ \frac{1}{2}\nabla_1^2\left[\left(\frac{1}{2}\nabla_1^2(\Phi^2) - \Phi\nabla_1^2\Phi\right)\frac{1}{k_{\mathrm{r}}^2} + \Phi^2\right](f^2)'. \qquad (20)$$

When $\Omega = 0$ (no rotation) the last two terms of (19) vanish, while the coefficients of the remaining terms in the linear and nonlinear parts of equations (19) and (20) change somewhat; however the general structure remains unaltered. When $\Omega = 0$ two extremely important properties of the equations become obvious. First, only the plane Laplacian operator acts on the terms with Φ^2 in the nonlinear parts of the equations, so leaving the dependence of Φ^2 on x and y unchanged. The relation between the coefficients of the terms having a single wave number does not change as a result of application of operator ∇_1^2. The relation between the coefficients of the terms having different horizontal wave numbers does not change. Therefore, for terms of type $\nabla_1^2(\Phi^2)$ to repeat the dependence of the linear terms Φ on x and y, they must have the same horizontal wave number as Φ.

The above arguments enable us to formulate a principle of recurrence: the nonlinear terms in the Boussinesq equations having the same horizontal wave number as the linear terms repeat the dependence of the functions (solutions) on the horizontal coordinate, since the relation between the coefficients of the nonlinear terms remains in this case unaltered.

The second property of equations (19) and (20) is that the non-linear terms are odd derivatives with respect to z of the quadratic linear terms. The appearance of nonlinear terms of second degree is of basic importance, since it is exactly they which produce asymmetry with respect to the vertical in the particular geometry of convective currents. In equations (19) and (20) such a form of asymmetry is introduced (as in /3/) by allowing for nonlinearity of the mean temperature profile.

Allowance for vorticity affects the structure of the equations in two ways: two additional components appear in the first brace of equation (19) and two additional terms in the last two braces. The group of terms entering the first brace does not violate the principle of recurrence as formulated on the basis of the properties of (17) and (18), while the following two terms have odd derivatives with respect to the horizontal coordinate and the conditions of recurrence (if we use only the properties of (17) and (18)) are not satisfied.

However, it is clear that the introduction of an additional condition

$$\Phi_x = \pm \frac{k_y}{k_x} \Phi_y \tag{21}$$

eliminates the two last braces, and equation (19) begins to satisfy the principle of recurrence formulated above.

Equation (19) also reveals other interesting properties: when Φ is a function only of x or y (corresponding to a two-dimensional convective current), the two last braces are eliminated and the principle of recurrence is conserved without the introduction of additional condition (21). It also follows from (19) that the effect of rotation shows up only when $\dfrac{\Omega^2}{v^2 k^4}$ is near unity, i. e., when the coefficient of turbulent viscosity $v = 10\ \mathrm{m}^2/\mathrm{sec}$ the rotation begins to influence perturbations with long waves of the order of several kilometers. Considering this and the constancy of Ω, all further reasoning applies only to processes of mesoscale convection.

Under free boundary conditions

$$T = w = u_z = v_z \Big|_{\substack{z=0 \\ z=h}} = 0 \tag{22}$$

equations (13) to (16) and conditions (18) and (21) possess the following solutions:

$$w = \sum_{p,q,s} A_{pqs}(t) \cos pk_x x \cos qk_y y \sin sk_z z -$$

$$- \widetilde{A}_{pqs}(t) \sin pk_x x \sin qk_y y \sin sk_z z,$$

$$u = \sum_{p,q,s} B_{pqs}(t) \sin pk_x x \cos qk_y y \cos sk_z z +$$

$$+ \widetilde{B}_{pqs}(t) \cos pk_x x \sin qk_y y \cos sk_z z,$$

$$v = \sum_{p,q,s} C_{pqs}(t) \cos pk_x x \sin qk_y y \cos sk_z z +$$

$$+ \widetilde{C}_{pqs}(t) \sin pk_x x \cos qk_y y \cos sk_z z,$$

$$T + T_2 = \sum_{p,q,s} D_{pqs}(t) \cos pk_x x \cos qk_y y \sin sk_z z -$$

$$- \widetilde{D}_{pqs}(t) \sin pk_x x \sin qk_y y \sin sk_z z. \tag{23}$$

Here A, B, C and D are time-dependent amplitudes, where $|A| = |\widetilde{A}|$, $|B| = |\widetilde{B}|$, etc.; p, q, s are positive integers; k_x, k_y, k_z are the wave numbers with respect to the corresponding axes.

Formal substitution of (23) into system (13) — (16) results in an infinite number of equations. The condition of recurrence allows one to choose the main part of the general spectrum of disturbances. Segel /9/ has shown that we have a one-parameter family of six interacting disturbances:

$$\begin{matrix} \cos \\ \sin \end{matrix} [\pi \alpha y],$$

$$\begin{matrix} \cos \\ \sin \end{matrix} \left[\frac{1}{2} \pi \alpha y\right] \times \begin{matrix} \cos \\ \sin \end{matrix} \left[\frac{\sqrt{3}}{2} \pi \alpha x\right], \tag{24}$$

for which the condition of recurrence is fulfilled. Relationship (24) implies the equality $k_x^2 = 3k_y^2$. We shall examine the partial, but representative case of two amplitudes, $A_{111}(t)$ and $A_{021}(t)$. System (13) — (16) then reduces to two ordinary nonlinear equations for the amplitudes (henceforth referred to as the amplitude equations). This approach entails an approximate solution of the problem, preserving its nonlinear character.

We do not wish to impose initial conditions since we are interested in steady solutions under given conditions, depending on time transitions from one such solution to another. We shall disregard any oscillatory solutions which may occur with the examined convection in a rotating fluid, because with the Prandtl number taken as unity the convective currents (in accordance with linear theory) appear primarily as steady convection, and with any further increase in the Taylor number they become oscillatory solutions.

In constructing the amplitude equations we shall limit ourselves to terms of third degree in the nonlinear parts of the equations, and in expansion (23) only the first three harmonics shall be taken into account.   The following procedure is used to derive the amplitude equations: from equations (14) — (16) amplitudes B_{pqs}, C_{pqs} and D_{pqs} are expressed in terms of A_{pqs} and from equation (13) amplitude A_{pqs} is expressed in terms of the fundamental amplitudes A_{111} and A_{021}. This procedure yields two ordinary nonlinear second-order differential equations (with respect to time).   The absolute vorticity Ω enters into the coefficients of the nonlinear terms with powers of 0 to 4.

Simplifications in the resulting amplitude equations are made as follows: terms with a second time derivative are omitted (inclusion of the second derivative in the examined problem adds two solutions, which are damped owing to their exponents); terms containing Ω to the second or higher degree and those terms having higher order of magnitude are excluded from the nonlinear terms of the equations. Consequently the amplitude equations assume the form

$$\dot{A}_{111} = EA_{111} - PA_{111}A_{021} - (Q + \Omega l_1)\, A_{111}^3 - (S - \Omega l_2)\, A_{111}A_{021}^2, \tag{25}$$

$$\dot{A}_{021} = EA_{021} - \left(\frac{1}{4}\, P + \Omega l_3\right) A_{111}^2 - T_{021}^3 - \frac{1}{2}\, (S - \Omega l_2)\, A_{111}^2 A_{021}, \tag{26}$$

where

$$E = \left[4g\alpha\beta k_y^2 - \nu^2 k^6 - \frac{F^2}{M} - \Omega^2 k_z^2\right] R,$$

$$F = 2g\alpha k_y^2 k_z \Gamma,$$

$$Q = \frac{1}{32}\left[\frac{1}{4}\,\frac{g\alpha\beta k_y^2 k_z^2}{\nu^2 k^2}\left(\frac{9}{k_y^2 + k_z^2} + \frac{1}{k_x^2 + k_z^2} + \frac{32}{k_z^2}\right) - \frac{k^2}{k_y^2 + k_z^2}\,\frac{G^2}{M} - \frac{1}{3}\,\frac{k^2}{k_x^2 + k_z^2}\,\frac{L^2}{N}\right] R,$$

$$G = 3k_z\left[\nu k^2\left(k_y^2 + k_z^2\right) + \frac{g\alpha\beta k_y^2}{\nu k^2}\right],$$

$$L = 3k_z\left[\nu k^2\left(k_x^2 + k_z^2\right) + \frac{g\alpha\beta k_y^2}{\nu k^2}\right],$$

$$M = 4g\alpha\beta k_y^2 - 64\nu^2\left(k_y^2 + k_z^2\right)^3 - 4\Omega^2 k_z^2,$$

$$N = 4g\alpha\beta k_y^2 - 64\nu^2\left(k_x^2 + k_z^2\right)^3 - 4\Omega^2 k_z^2,$$

$$P = \frac{3}{4}\,\frac{k_z^2}{k_y^2 + k_z^2}\,\frac{FG}{M}\, R,$$

$$R = (2 \nu k^4)^{-1},$$

$$S = 4Q - T,$$

$$T = \frac{1}{2} \frac{g \alpha \beta k_y^2}{\nu^2 k^2} R,$$

$$l_1 = \frac{\sqrt{3}}{64} \frac{g \alpha \beta k_y^2 k_z^2}{\nu^3 k^4} \left(\frac{3}{k_y^2 + k_z^2} - \frac{1}{k_x^2 + k_z^2} \right) R,$$

$$l_2 = \frac{\sqrt{3}}{32} \frac{g \alpha \beta k_y^2 k_z^2}{\nu^3 k^4} \left(\frac{9}{k_y^2 + k_z^2} - \frac{1}{k_x^2 + k_z^2} \right) R,$$

$$l_3 = \frac{3}{16} \frac{k_z^2}{k_y^2 + k_z^2} \frac{F}{M} \left\{ \frac{2\sqrt{3} \, k_z}{k^2} \left[(k_y^2 + k_z^2)(2k_y^2 - k_z^2) + \right. \right.$$

$$\left. \left. + \frac{g \alpha \beta k_y^2}{2\nu^2 k^2} \right] + 3\sqrt{3} \, k_y^2 k_z \right\} R.$$

ANALYSIS OF THE EQUATIONS

If terms with time derivatives and nonlinear components are disregarded in equations (25) and (26), we derive the same classical relationship as for a linear temperature profile /7/:

$$\mathrm{Ra}_{cr} \to 3\pi^4 \left(\frac{T}{2\pi^4} \right)^{\frac{2}{3}}, \tag{27}$$

where $T = \dfrac{\Omega^2 h^4}{\nu^2}$ is the Taylor number and h the depth of the convective layer. It also follows from (25) and (26) that consideration
of the linear terms alone results in amplitudes A_{111} and A_{021} changing
in like manner with time, while the finite form of currents remains
undetermined.

If in equations (25) and (26) we set $\Omega = 0$ they will become identical to those obtained in /3/, and the form of the convective currents
will be determined entirely by the mean temperature profile.

Analysis of the complete equations (25) and (26) is begun with an
examination of the equilibrium solution:

I. $A_{111} = A_{021} = 0,$

IIa, b. $A_{111} = 0, \quad A_{021} = \pm (E/T)^{\frac{1}{2}},$

IIIa, b. $A_{111} = 2A_{021}, \quad A_{021} = (8Q - T + 8\Omega l_1)^{-1} \times$

$$\times \left[-\frac{P + \Omega l_3}{2} \pm \sqrt{\left(\frac{P + \Omega l_3}{2} \right)^2 + E(8Q - T + 8\Omega l_1)} \right],$$

IVa, b. $A_{111} = -2A_{021},$ \hfill (28)

$$A_{021} = (8Q - T + 8\Omega l_1)^{-1}\left[-\frac{P + \Omega l_3}{2} \pm \right.$$
$$\left. \pm \sqrt{\left(\frac{P + \Omega l_3}{2}\right)^2 + E(8Q - T + 8\Omega l_1)}\right]. \tag{28}$$

Case I represents the absence of currents, case II two-dimensional cells (billows), cases IIIa and IVa hexagonal cells with ascending currents in the center, while IIIb and IVb represent hexagonal cells with descending currents in the center.

An accurate analytical solution of equations (25) and (26) cannot be obtained, so we carried out an approximate linearized analysis of the stability of the equilibrium solutions in order to classify ther as nodes (stable or unstable), saddle points, etc. The existence of stable nodes /1/ determines a predominance of geometrical perturbations.

The amplitudes are expressed in the form

$$A_{111} = x_0 + x,$$
$$A_{021} = y_0 + y, \tag{29}$$

where x_0 and y_0 are the equilibrium solutions of (28). Substitution of (29) into (25) and (26) and linearization yield

$$x' = \{E - Py_0 - 3(Q - \Omega l_1)x_0^2 - [4Q - T + \Omega(4l_1 - l_2)]y_0^2\}x +$$
$$+ \{-Px_0 - [8Q - 2T + \Omega(8l_1 - 2l_2)]x_0 y_0\}y, \tag{30}$$
$$y' = \left\{-2\left(\frac{1}{4}P + \Omega l_3\right)x_0 - [4Q - T + \Omega(4l_1 - l_2)]x_0 y_0\right\}x +$$
$$+ \left\{E - 3Ty_0 - \frac{1}{2}[4Q - T + \Omega(4l_1 - l_2)]x_0^2\right\}y. \tag{31}$$

Equations (30) and (31) can be written as

$$\frac{\partial x}{\partial y} = \frac{\alpha y + \beta x}{\eta y + \delta x}, \tag{32}$$

where

$$\alpha = -Px_0 - [8Q - 2T + \Omega(8l_1 - 2l_2)]x_0 y_0,$$
$$\beta = E - Py_0 - 3(Q + \Omega l_1)x_0^2 - [4Q - T + \Omega(4l_1 - l_2)]y_0^2,$$
$$\delta = -2\left(\frac{1}{4}P + \Omega l_3\right)x_0 - [4Q - T + \Omega(4l_1 - l_2)]x_0 y_0,$$
$$\eta = E - 3Ty_0 - \frac{1}{2}[4Q - T + \Omega(4l_1 - l_2)]x_0^2. \tag{33}$$

The solution of (32) represents a saddle point when $\alpha\delta - \beta\eta > 0$ and a node when $\alpha\delta - \beta\eta < 0$. The node is stable when $\beta + \eta < 0$ and unstable when $\beta + \eta > 0$.

A general calculation leads to very unwieldy expressions. More tangible results are obtained in the particular case when rotation is disregarded in the equilibrium solutions of (28). Then, substitution of solution (28) in (33) gives conditions under which a certain type of perturbation is stable. Some idea of the transition from one stable form of convection to another can be obtained if we introduce parameter $\zeta = PE^{-\frac{1}{2}}$. Since coefficient E assumes a high positive value, the sign of ζ is determined by the sign of factor $\dfrac{F}{M}$. When Ra < < 16 Ra $_{cr}$, $M < 0$. The sign of F depends on the sign of coefficient Γ. As we are interested in the behavior of the fluid near the critical Rayleigh number, the case $M < 0$ is treated below. In these conditions ζ will become positive when $\Gamma < 0$, i. e., when the temperature gradient decreases with height. With an increase in the temperature gradient with height, ζ will become negative.

Figure 1 depicts the results of the stability analysis for perturbations of different geometry. The center of this graph corresponds to a linear temperature profile without rotation. Values of the absolute vorticity Ω are plotted along the abscissa, and of parameter ζ along the ordinate. The curves in the figure signify the limits of the stable nodes for hexagonal cells, while straight lines indicate the limits of the stable nodes for two-dimensional billows. The equations of these limits are

$$\text{I.} \quad \zeta = \pm \left\{ \frac{(4Q - 2T)^2 - 12\Omega(4Q - 2T)\left[\dfrac{16l_1(2Q - T) - l_2(8Q - 3T)}{8Q - T}\right]}{(4Q + T) - \Omega(a_1 + a_2)} \right\}^{\frac{1}{2}},$$

where

$$a_1 = 4\sqrt{3}\,\frac{Qk_z}{G}\left\{ \frac{2}{k^2}\left[(k_y^2 + k_z^2)(2k_y^2 - k_z^2) + \frac{g\alpha\beta k_y^2}{2\nu^2 k^2}\right] + 3k_y^2 \right\},$$

$$a_2 = -\,\frac{6\,[16l_1(4Q + T) - l_2(8Q + 3T)]}{8Q - T}.$$

$$\text{II.} \quad \zeta = \pm \left[\frac{(8Q + 2T)^2 + \Omega(8Q + 2T)(16l_1 - 3l_2)}{(16Q + T) + \Omega(16l_1 - 3l_2)} \right]^{\frac{1}{2}}.$$

$$\text{III.} \quad \zeta = \pm\,\frac{4Q - 2T - \Omega l_2}{\sqrt{T}}.$$

Dashes in Figure 1 outline the area where stable two-dimensional billows occur, while the hatched areas show where stable hexagonal cells occur. On the line $\Omega = 0$, the values of ζ at which transition from node to saddle point occurs have the identical values as those obtained in /3/* $\left(d = \dfrac{4Q - 2T}{\sqrt{4Q + T}}, \quad e = \dfrac{4Q - 2T}{\sqrt{T}} \right)$. The disposition of the curves leads to the following conclusions.

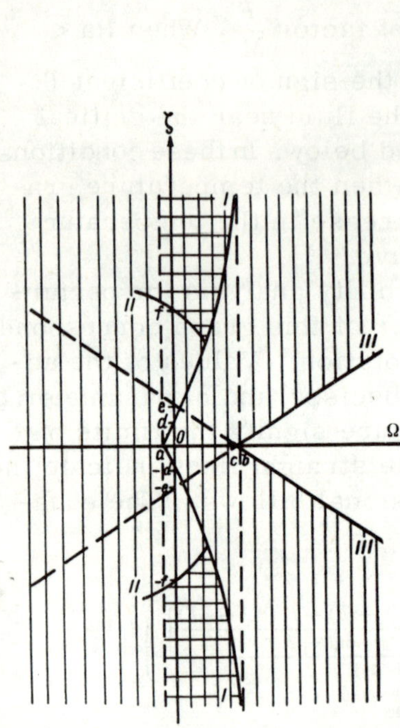

FIGURE 1. Dependence of the stability of convective cells of different forms on ζ and Ω.

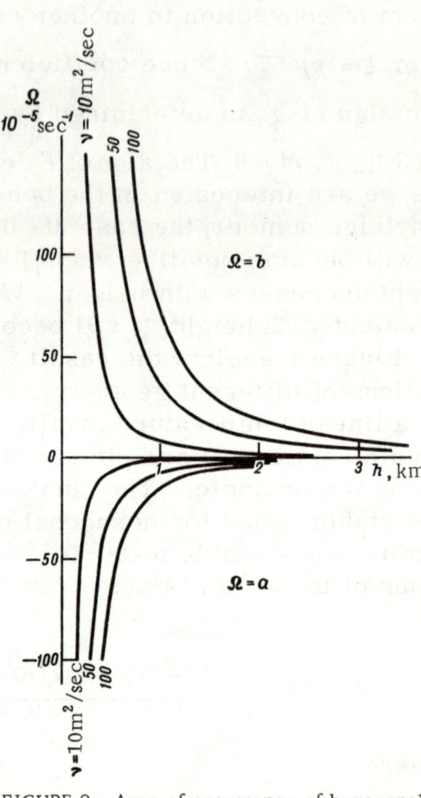

FIGURE 2. Area of occurrence of hexagonal convective cells as a function of the thickness of the convective layer h, absolute vorticity Ω, and coefficient of turbulent viscosity ν:

$$\Omega = b = 0.75\nu \frac{\pi^2}{h^2}, \quad \Omega = a = -0.16\nu \frac{\pi^2}{h^2}.$$

1.   When $\Omega < c = (4Q - 2T)/l_2$, convective cells in the form of two-dimensional billows are stable forms of currents, independently of the type of temperature stratification.   When $\Omega > c$, a current in the form of two-dimensional billows is possible only with the occurrence of nonlinear temperature profiles, and with increase in Ω the nonlinearity must also increase.

* The limit of stable nodes for cases III and IV in this paper should be obtained for
$$\zeta = f = \pm \frac{8Q + 2T}{\sqrt{16Q + T}}.$$

2.   Currents in the form of hexagonal convective cells can be stable only when $a < \Omega < b$

$$\left(a = \frac{(4Q - 2T)(8Q - T)}{12\left[16l_1(2Q - T) - l_2(8Q - 3T)\right]}, \quad b = \frac{4Q + T}{a_1 + a_2} \right),$$

i. e., with increase in the absolute value of the vorticity the hexagonal convective cells cease to be stable forms of convection and must disintegrate.

3.   Within the interval $a < \Omega < b$ hexagonal convective cells with descending currents in their center can form only with a temperature gradient decreasing with height, while cells with ascending currents in the center can only arise with a temperature gradient increasing with height.

4.   With increasing curvature of the temperature profile, there is an increase in the critical value of the positive absolute vorticity at which hexagonal cells and two-dimensional billows can still form.

5.   An area of ζ and Ω exists where the occurrence of two-dimensional billows and hexagonal cells is little probable.

6.   With near-linear temperature stratification and rather large positive vorticity $(\Omega > c)$ neither two-dimensional billows nor hexagonal cells can form.

Thus, analysis of amplitude equations (25) and (26) shows that rotation, after the absolute vorticity reaches a certain value, breaks up three-dimensional convective cells, leaving in a stable state only those convective currents which exist in the form of two-dimensional billows.

It is of interest to compute the limiting value of the absolute vorticity at which hexagonal cells can no longer form.   If we take into account that the horizontal dimensions of convective cells in the atmosphere are somewhat larger than the vertical dimensions, the general wave number can be represented in the form $k^2 \approx \dfrac{\pi^2}{h^2}$, where h is the thickness of the convective layer.   Then, the equation of the asymptotes to hyperbola I assumes the form $\Omega \simeq 0.75 v \pi^2 / h^2$, while the limit of the negative values of vorticity (a) is $\Omega \simeq -0.16 v \pi^2 / h^2$.

The dependence of Ω_{cr} on h and v for both limits is shown in Figure 2.   The space between the two hyperbolas lying on both sides of the abscissa and corresponding to the same viscosity indicates the area of Ω and h where hexagonal convective cells may occur. It is seen that this area broadens with increasing turbulent viscosity.   Another interesting point is that smaller hexagonal cells can form with larger absolute vorticity.

The limiting value of Ω_{cr} at which two-dimensional billows cease to form (with linear temperature stratification) is also a function of viscosity and the scale of the perturbation, $\Omega \simeq 0.72 v\pi^2/h^2$. Smaller two-dimensional billows can form with larger positive values of absolute vorticity.

COMPARISON WITH EMPIRICAL DATA

The above scheme is in agreement with quasihexagonal cells observed by satellites over regions with anticyclonic or small cyclonic vortices /2, 5/. Our analysis of the amplitude equations also explains the previously mentioned absence of very large hexagonal convective cells in the atmosphere. Cells with diameters exceeding 100 km are hardly ever encountered in the atmosphere /2, 5/. If we take into account that the ratio between the diameter of the cells and the thickness of the convective layer is on the average 100, Figure 2 implies that for the break-up of cells with a diameter of 30 km it is sufficient that $-0.17 \cdot 10^{-5} \sec^{-1} \geqslant \Omega \geqslant 0.8 \cdot 10^{-5} \sec^{-1}$, i.e., that the value of the absolute vorticity is quite small. If we use average values of the thickness of the convection layer for cells /3/, we find that for $v = 10 \text{ m}^2/\sec$ cells of $30 - 40$ km diameter ($h = 1.6$ km) can exist with absolute vorticity $\Omega \simeq 3 \cdot 10^{-5} \sec^{-1}$, which is very near actual conditions in the atmosphere.

BIBLIOGRAPHY

1. Andronov, A.A., A.A.Vitt, and S.E.Khaikin. Teoriya kolebanii (Theory of Oscillations). Moskva, Fizmatgiz. 1959.
2. Vel'tishchev, N.F. and L.I.Silaeva. Konvektivnye oblachnye yacheiki po nablyudeniyam s iskusstvennykh sputnikov Zemli (Convective Cloud Cells as Observed from Artificial Earth Satellites). — Trudy Gidromettsentra SSSR, No.20. 1968.
3. Vel'tishchev, N.F. Yacheikovaya konvektsiya v atmosfere (Cellular Convection in the Atmosphere). — Trudy Gidromettsentra SSSR, No.50. 1969.
4. Vorob'eva, E.F. and B.N.Trubnikov. Yacheikovaya konvektsiya pri anizotropnoi turbulentnosti i nalichii polya otnositel'nogo vikhrya (Cellular Convection in the Presence of Anisotropic Turbulence and a Relative Vorticity Field). — Trudy TsAO, No.75. 1967.
5. Zamorskii, A.D. and L.S.Minina. Yacheistye sistemy oblakov po dannym meteorologicheskikh sputnikov Zemli (Cellular Cloud Systems According to Meteorological Earth Satellite Data). — Trudy Gidromettsentra SSSR, No.26. 1968.
6. Trubnikov, B.N. Nekotorye voprosy teorii svobodnoi (yacheikovoi) i vynuzhdennoi konvektsii (Some Problems in the Theory of Free (Cellular) and Forced Convection). — Trudy TsAO, No.75. 1967.
7. Chandrasekhar, S. Hydrodynamic and Hydromagnetic Stability. Oxford, Clarendon. 1961.
8. Krishnamurti, R. Finite Amplitude Convection with Changing Mean Temperature. —J. Fluid Mech., Vol.33, Part 1. 1968.
9. Segel, L.A. The Nonlinear Interaction of a Finite Number of Disturbances to a Layer of Fluid Heated from Below. J. Fluid Mech., Vol.21, Part 2. 1965.

UDC 551.509.311:551.515

SOME PARAMETERS OF CYCLONIC CLOUD VORTICES

T. P. Popova

This paper deals with the connections between morphological features of cloud vortices and the position of isallobaric centers of depressions and the relations between cloud forms and thermal advection. It is shown that a uniform cloud cover corresponds to a typical vertical advection profile, whereas cloudiness of a mild cumuliform and stratiform type indicates a complex profile of thermal advection.

Cyclonic cloud vortices are distinguished by a large variety of shapes and dimensions. But in every cloud vortex one can distinguish basic cloud spirals represented by a cloud band of from 100 to 400 km width whose curvature gradually decreases with distance from the center of the cloud vortex.

An analysis of 120 cyclonic cloud vortices, chosen from a series of observations from April 1966 through December 1967 over Europe and West Siberia, shows that in cloud vortices from one to several spirals can be observed. Morphological indicators can be used to separate three fundamental types of cyclonic cloud vortices.

The first type consists of cloud vortices having cloud spirals with straight extensions similar to hyperbolic or logarithmic spirals (Figure 1, curves I and II). The linear dimensions of the cloud vortices usually exceed 1000 km. One, sometimes two, but seldom three, such spirals may belong to one such cloud vortex. Such cloud vortices correspond to thermally asymmetric, developed, cut-off and mainly mobile depressions.

The second type of cloud vortices consists of those having a regular, circular shape with dimensions of 500 to 800 km (Figure 1, curve III), comprising one or several symmetrically arranged cloud spirals recalling an Archimedian spiral. These vortices are connected with cut-off, thermally symmetric quasistationary depressions.

The third type are cloud vortices of irregular, odd shapes (Figure 1, curve IV). These vortices may have any dimensions from some hundreds to 1500—2000 km. Irregularly formed vortices appear in cyclone systems with a complex structure of the pressure and temperature field. These depressions usually develop as a result of regeneration or merging of two or several

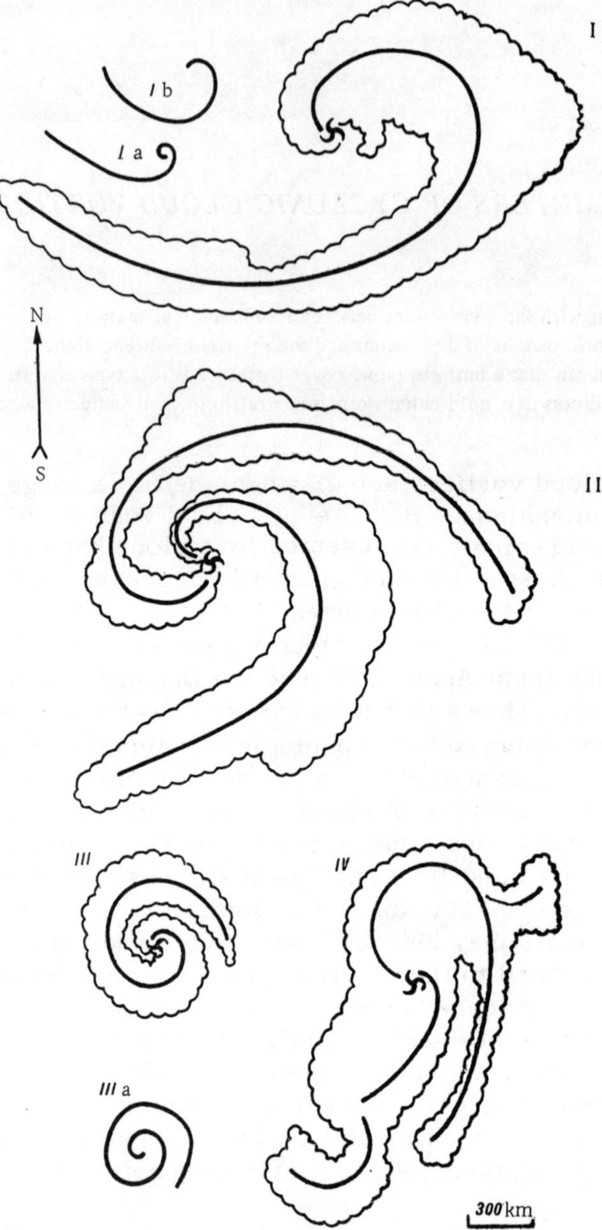

FIGURE 1. Types of cloud vortices:

I — cloud vortex on 19 May 1966; Ia — hyperbolic spiral; Ib — logarithmic spiral (first type); III — cloud vortex with two spirals on 23 June 1966; III — cloud vortex on 7 July 1967; IIIa — Archimedian spiral (second type); IV — cloud vortex on 7 April 1966 (third type).

depressions, and also as a result of the appearance of new depressions and wave perturbations within the system of the main depression.

Every cloud vortex undergoes a well-defined life cycle during which the form of its cloud spirals changes; therefore, subdivisions of cloud vortices into the above types are conditional. The character of the clouds from which cloud vortices are formed also changes with time. At the initial stage of the appearance of a cyclonic vortex the latter consists of predominantly stratiform clouds which, with time, evolve into cumuliform clouds.

From a study of 120 cloud vortices the first type proved to be the most frequent. There were 60 elongated cloud vortices with one spiral, and 28 with two or more spirals. There were only 12 circular and 20 complex and irregular vortices.

In order to obtain average data for the linear dimensions of cloud vortices we made an additional selection from 148 vortices for 1968. A statistical analysis of 268 cloud-vortex situations was carried out. These situations consisted of 76 cyclonic cloud vortices, each of which persisted for several days. It was seen that cloud vortices having linear dimensions of 600 — 1500 km are the most frequent. If each cloud vortex is inserted into a rectangle, so that each side of the rectangle is parallel to the direction of movement of the depression center, the length of this side can reach 2500 km or more with some vortices, and that of the smaller side 1800 km. Table 1 presents the frequency distribution of horizontal dimensions of cloud vortices. It is seen that between the long sides of the rectangle and its short sides (i. e., between the geometric dimensions of the cloud vortices) there is a well defined proportion. The long side of the vortex does not usually exceed double the length of the short side.

The average life cycle of a cloud vortex is about $3\frac{1}{2}$ days. The most frequently observed cloud vortices lasted 2 or 3 days. At the end of this period the cloud cover usually becomes so distorted that its spiral structure becomes difficult to discern. The great majority of cloud vortices were associated with well developed depressions. Some average characteristics of these depressions are the velocity of displacement and the change of pressure at their center, as shown below (the velocity was rounded off to the nearest 10 km/hr and the pressure to 1 millibar):

Velocity (km/hour)	0	10	20	30	40	50	60	70
Number of cases	7	9	27	18	18	3	4	2

Change of pressure (mb per 12 hours)	+10	+9	+8	+7	+6	+5	+4	+3	+2	+1
Number of cases	1	3	2	4	—	4	6	5	12	9
Change of pressure (mb per 12 hours)	0	−1	−2	−3	−4	−5	−6	−7		
Number of cases	14	21	10	3	—	3	1	2		

Pressure at the center of the depressions with cloud vortices was generally (80%) between 980 and 1010 mb.

The nomogram (Figure 2) shows the position of cloud vortices relative to the pairs of isallobaric centers of depression. The nomogram is constructed from a relative system of coordinates; F is the center of pressure fall, R the center of pressure rise, RF the line connecting centers of pressure rises and falls, B the projection of the vortex center on the line RF, C the distance between F and R, b the distance of the cloud vortex from the line RF (positive if the cloud vortex lies to the left of RF and negative to the right), a the distance from the center of pressure rise (R) to the projection of the cloud vortex on the line RF. The distance RF was taken as unity.

The zero point of the diagram was placed at R. As abscissa we selected ratio $\dfrac{a}{C}$ and as ordinate ratio $\dfrac{b}{C}$. Averaging was carried out over some area, the dimensions of which are shown in the lower right corner of the graph.

The isolines used in the diagram show the frequency of occurrence of the position of the cloud vortices relative to the pairs of isallobaric centers.

It follows from the nomogram that in the great majority of cases (76 out of 88, or 86%) cloud vortices lie between centers of fall and rise of pressure, while somewhat more than half (58%) lie to the left of this line and somewhat less than half to the right.

The nomogram also shows that in spite of some dispersion of points about line RF the vortices are mainly concentrated around this line: 62% of all cloud vortices lie at a distance of $\frac{1}{4}$ RF to the right and left of it and 86% at a distance of $\frac{1}{2}$ RF.

Thus, as a rule, the cloud vortex lies in a square whose side is equal to the distance between the centers of pressure and rise. The line RF represents the center line of this square.

The intensity of the isallobaric centers, i. e., the total pressure tendency in the middle of the fall and rise centers, usually does not exceed 1 mb (73%). This shows the small change of pressure taking place within the depth of the depression with which the cloud vortex is connected. The mean total pressure tendency at the fall and rise centers is 0.1 mb, i. e., practically zero. The standard deviation $\sigma = \pm 1.3$ mb.

TABLE 1. Frequency (No. of cases) of horizontal dimensions of cyclonic cloud vortices ($n = 268$)

Axis	Horizontal dimensions (×100 km)							
	3	4.5	6	7.5	9	10.5	12	13.5
Large	2	5	8	13	28	31	30	34
Small	7	12	36	38	54	40	30	17
Axis	Horizontal dimensions (×100 km)							
	15	16.5	18	19.5	21	22.5	24	25.5
Large	24	23	22	16	5	9	10	8
Small	18	11	3	2				

TABLE 2. Position of isallobaric center pairs relative to cloud vortices with one spiral (No. of cases)

Pressure rise	Pressure fall	
	outside and in front of the spiral	inside the spiral
Outside and at the rear of the spiral	5	32
Inside the spiral	13	10

TABLE 3. Position of isallobaric center pairs relative to cloud vortices with two (three) spirals (No. of cases)

Pressure rise	Pressure fall	
	at the main spiral or near the center of the vortex	between spirals
Outside and at the rear of the spiral	13	2
In the second (third) spiral	6	7

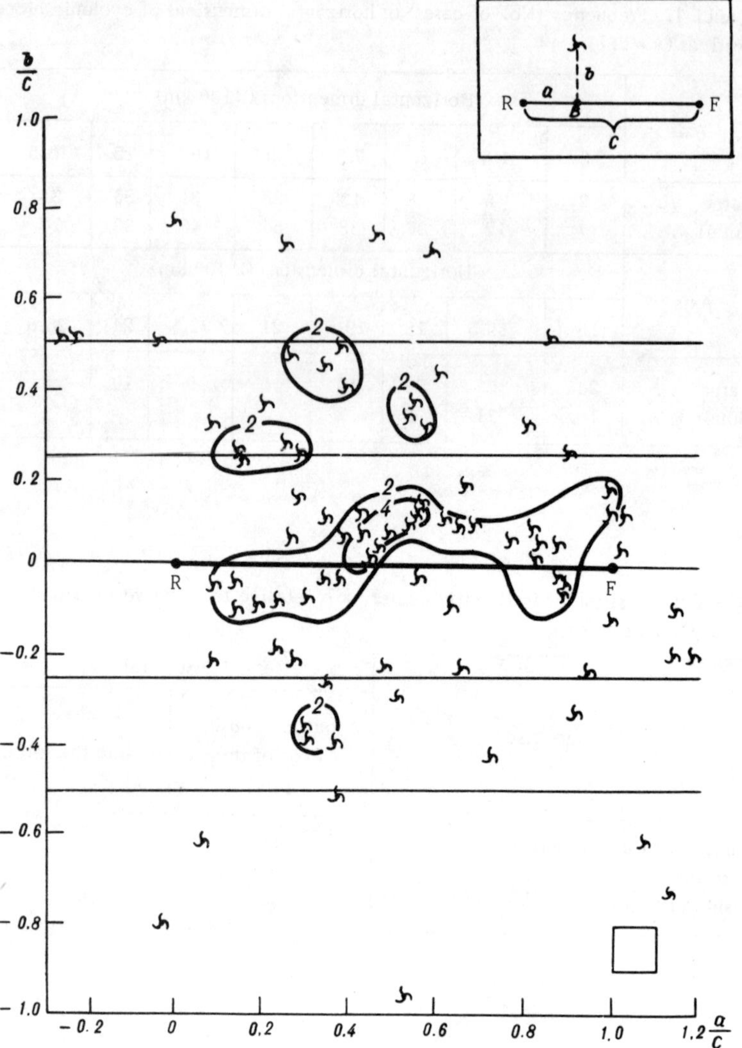

FIGURE 2.   Position of cloud vortices relative to the isallobaric centers of depression.   The diagram in the upper right corner defines the notation: R is the center of pressure rise, F the center of pressure fall. The averaging area is shown at the lower right of the graph.

In all vortices a well defined regularity was observed in the position of the isallobaric center pairs relative to the cloud spirals. Data on this subject are presented for the first type of spirals (elongated) in Tables 2 and 3.

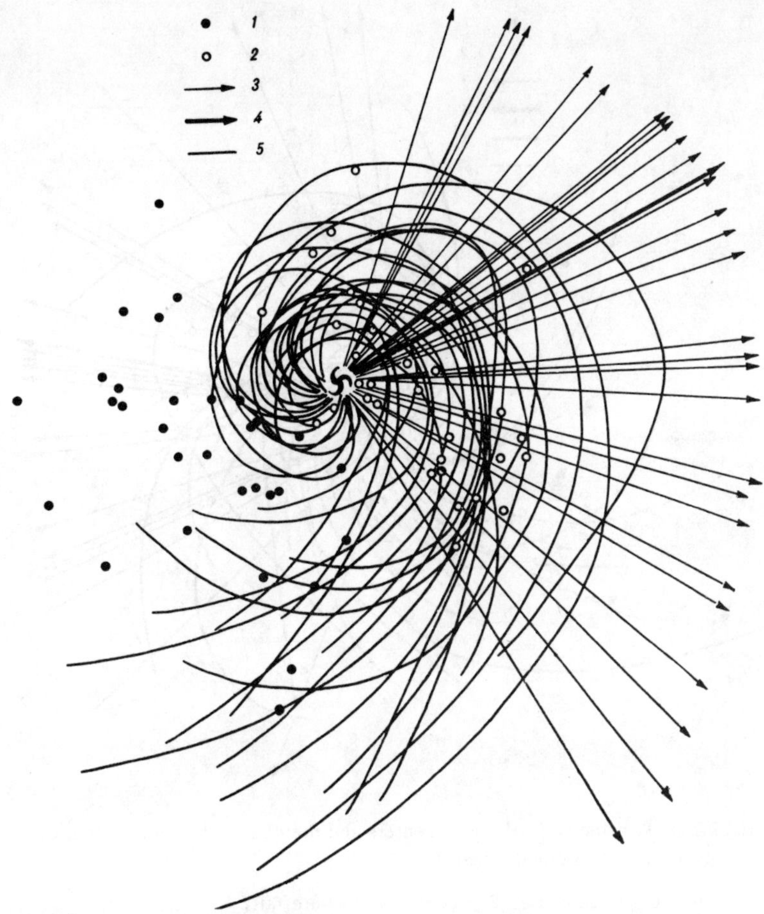

FIGURE 3. Position of isallobaric centers and the direction of movement of depressions with one cloud spiral:

1 — center of pressure rise: 2 — center of pressure fall; 3 — direction of movement of individual depressions; 4 — average direction of movement of all depressions; 5 — cloud spiral.

In the case of cloud vortices with one spiral (Figure 3) the fall center lies most often near the head of the spiral where its curvature is most pronounced. The rise center lies in the rear, outside the spiral or in those parts where the curvature is already small, and more seldom near the center of the vortex. We found no cases in which the fall center lay at the rear and the rise center in front of the spiral.

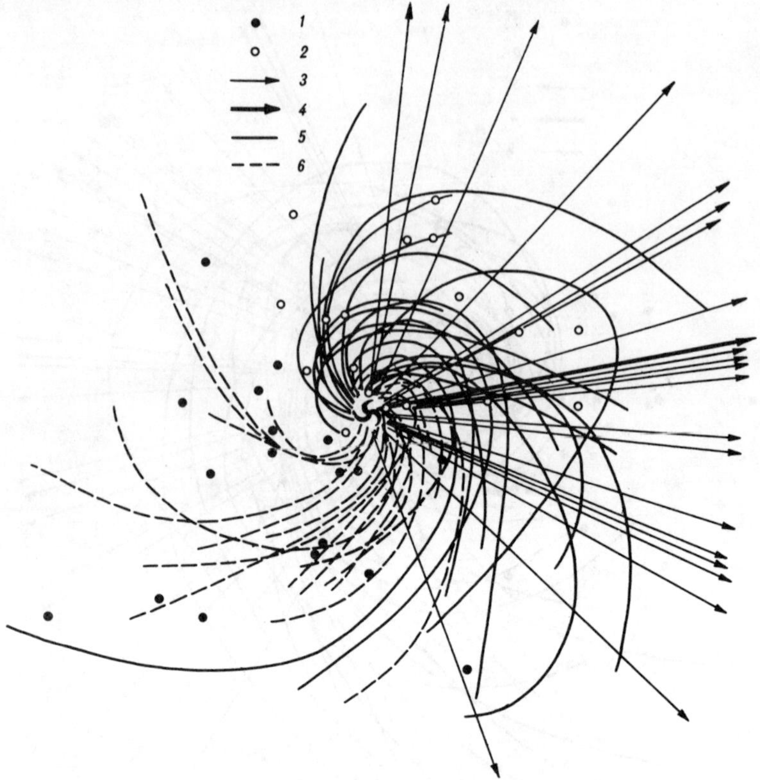

FIGURE 4. Position of isallobaric centers and the direction of movement of depressions with two cloud spirals:

1 — center of pressure rise; 2 — center of pressure fall; 3 — direction of movement of individual depressions; 4 — average direction of movement of all depressions; 5 — first cloud spiral; 6 — second cloud spiral.

With cloud vortices having two spirals (Figure 4) the fall center is observed at the main spiral or between the spirals, while the rise center lies on or behind the second spiral.

The direction of movement of depressions with which cloud vortices were associated was in agreement with the position of the pair of isallobaric centers in the cloud vortex system.

Depressions whose cloud vortex had one spiral moved in a regular path parallel to those parts of the spiral which had the least curvature (the tail of the spiral).

When cloud vortices have two or three spirals the corresponding depression moves almost parallel to those parts of the second or third spiral having the least curvature, and frequently perpendicular to that part of the first spiral with the least curvature, i.e., perpendicular to the direction of movement of the depressions having one spiral.

Circular and irregularly shaped cloud vortices do not behave according to the same rules as depressions having cloud spirals, as
regards either their movement or the position of their isallobaric
center pairs (in relation to the cloud spirals).

An important parameter of state of the atmosphere is thermal
advection, considered as the transport of thermally nonuniform air.
As known, the types and amount of clouds are closely associated with
the sign of thermal advection. Here we are interested in the quantitative relations between cloudiness as determined by satellite data
and thermal advection. To obtain such relationships we analyzed
simultaneously TV pictures of the Meteor and ESSA satellites and
values of thermal advection computed in the form of the Jacobian of
(T, P) and (T, H) at grid points of a uniform network with steps of
150 km. Photographs of cloud vortices were chosen for this analysis. For the determination of cloud types the photographs were
divided into two basic forms: cumuliform and stratiform. Stratocumulus was treated separately, including clouds of both stratiform
and cumulus types.

Results of statistical processing for cases of cloudiness determined by Meteor satellite are presented in Table 4. An analogous
analysis for cloudiness determined by ESSA satellite is given in
Table 5.

Before analyzing the results it should be noted that, nearly everywhere, the number of cases at the earth's surface and at the 300-mb
level is considerably less than for the middle troposphere. The
reason is that unreliable data were discarded during analysis, and it
was seen that more such data were found at the limits of the troposphere than in its middle.

Data listed in Table 4 and similar data yielded some very interesting conclusions. Stratified clouds occur relatively seldom in
small amounts (less than 50% of the area within a 150×150 km
square covered by clouds). Stratiform clouds usually cover more
than 60% of the earth's surface, as is evident from the number of
cases of both cloud amount gradations.

Clouds of the stratocumulus type also mostly cover a considerable part of the earth's surface. In contrast, cumuliform clouds
are more frequently found in smaller quantities, i. e., the area free
of these clouds was more frequently larger than the area covered
by them.

Cloudiness including at the same time stratiform and cumuliform
types occupies considerably larger areas: on the basis of the number of cases actually investigated they occupy 70%, while stratus
occupies about 10% and cumuliform clouds about 20%.

TABLE 4. Mean values of thermal advection with different cloud types, determined according to Meteor TV photographs

P, mb	$\overline{A}$	$\sigma_{\overline{A}}$	$\rho_{\overline{A}}$	σ_A	ρ_A	n
No clouds						
Surface	0.42	0.32	0.21	3.77	2.52	147
850	0.30	0.14	0.09	2.02	1.35	235
700	—0.02	0.14	0.09	2.22	1.49	235
500	0.17	0.17	0.11	2.74	1.84	235
300	—1.02	0.32	0.21	4.11	2.75	147
Stratiform clouds (10 — 50%)						
Surface	3.80	0.72	0.48	1.62	1.08	5
850	0.29	0.72	0.48	2.17	1.45	9
700	0.64	0.75	0.50	2.26	1.51	9
500	1.70	0.91	0.61	2.74	1.83	9
300	—0.20	0.33	0.22	0.99	0.66	5
Stratiform clouds (60 — 100%)						
Surface	1.41	0.73	0.49	4.25	2.85	55
850	2.19	0.42	0.28	3.59	2.40	68
700	1.13	0.20	0.13	1.69	1.13	68
500	0.60	0.20	0.13	1.62	1.08	68
300	0.47	0.42	0.28	2.92	1.96	55
Cumulus (10 — 50%)						
Surface	0.41	0.32	0.21	3.27	2.19	102
850	0.19	0.17	0.11	1.90	1.27	123
700	—0.20	0.20	0.14	2.36	1.58	123
500	—0.21	0.22	0.15	2.54	1.70	123
300	—0.33	0.28	0.19	3.08	2.06	102
Cumulus (60 — 100%)						
Surface	—1.54	0.65	0.44	3.48	2.33	69
850	—0.63	0.33	0.22	2.40	1.61	89
700	—0.38	0.24	0.16	1.77	1.18	89
500	—1.26	0.40	0.27	2.84	1.90	89
300	—2.26	0.73	0.49	4.81	3.22	69
Stratocumulus type I (10 — 50%)						
Surface	—0.79	0.66	0.44	2.38	1.59	15
850	—0.35	0.28	0.19	1.32	0.88	15
700	0.13	0.42	0.28	1.95	1.31	15
500	2.26	0.32	0.21	1.43	0.96	15
300	1.83	0.55	0.37	2.06	1.38	15

TABLE 4 (Continued)

P, mb	$\overline{A}$	$\sigma_{\overline{A}}$	$\rho_{\overline{A}}$	σ_A	ρ_A	n
Stratocumulus type I (60 — 100%)						
Surface	—1.95	0.36	0.24	2.87	1.92	96
850	—1.09	0.20	0.13	2.03	1.36	96
700	—0.24	0.14	0.09	1.53	1.02	96
500	1.06	0.17	0.11	1.66	1.11	96
300	0.76	0.35	0.23	3.00	2.01	96
Stratocumulus type II (10 — 50%)						
Surface	1.39	1.18	0.79	4.27	2.86	18
850	1.65	0.47	0.31	2.26	1.51	33
700	—0.20	0.42	0.28	2.05	1.37	23
500	—1.17	0.56	0.38	2.68	1.80	23
300	—0.60	0.52	0.35	2.26	1.51	18
Stratocumulus type II (60 — 100%)						
Surface	0.69	0.36	0.24	3.10	2.08	97
850	0.95	0.14	0.09	1.98	1.33	169
700	—0.13	0.14	0.09	1.76	1.18	169
500	—1.10	0.17	0.11	2.37	1.59	169
300	—0.18	0.28	0.19	2.65	1.78	97
Stratocumulus and Cumulus type I (10 — 50%)						
Surface	—0.36	1.03	0.69	2.91	1.95	10
850	—0.88	0.33	0.22	1.34	0.90	16
700	—0.33	0.55	0.37	2.21	1.48	16
500	—1.70	0.97	0.65	3.88	2.60	16
300	0.52	0.88	0.59	2.65	1.78	10
Stratocumulus and Cumulus type I (60 — 100%)						
Surface	—0.73	0.56	0.38	3.56	2.38	45
850	—0.99	0.14	0.09	1.30	0.87	80
700	—0.01	0.17	0.11	1.62	1.08	80
500	0.87	2.26	0.17	2.40	1.61	80
300	0.06	0.75	0.50	4.74	3.18	45
Stratocumulus and Cumulus type II (10 — 50%)						
Surface	—0.37	0.66	0.44	2.64	1.77	17
850	0.34	0.37	0.25	1.67	1.12	19
700	—1.14	0.49	0.33	2.20	1.47	19
500	—1.78	0.51	0.34	2.26	1.51	19
300	—0.35	0.42	0.28	1.64	1.10	17
Stratocumulus and Cumulus type II (60 — 100%)						
Surface	1.45	0.74	0.50	4.06	2.72	33
850	—0.01	0.14	0.09	1.06	0.71	35
700	—0.61	0.20	0.13	1.57	1.05	35
500	—1.87	0.20	0.13	1.67	1.12	35
300	—0.12	0.56	0.38	3.10	2.08	33

Note. $\overline{A}$ is advection in deg/hr·10; $\sigma_{\overline{A}}$ and $\rho_{\overline{A}}$ are mean square and mean probable deviation from mean value of advection; σ_A and ρ_A are mean square and mean probable deviation of individual advection values.

TABLE 5. Mean values of thermal advection in the presence of different cloud types, determined according to ESSA TV photograph satellites

Cloud type	Advection (deg/hr·10)				
	surface	850 mb	700 mb	500 mb	300 mb
No clouds, $n=369$ and 477	−0.13	−0.44	−0.24	−0.11	−0.68
Stratiform (10 − 50%), $n=2$ and 7	0.95	0.34	0.21	−0.30	−0.10
Stratiform (60−100%), $n=82$ and 171	0.43	1.93	1.42	0.88	1.54
Cumuliform (10−50%), $n=229$ and 250	0.0	−1.01	−0.08	−0.12	−0.25
Cumuliform (60−100%), $n=64$ and 80	−0.22	0.0	−0.21	0.18	0.18
Stratocumulus type I (10−50%), $n=12$ and 34	−0.41	−1.07	0.57	0.63	0.08
Stratocumulus type I (60−100%), $n=98$ and 360	−0.09	−0.96	0.78	0.83	0.34
Stratocumulus type II (10−50%), $n=17$ and 53	−0.17	1.62	0.54	−0.44	0.41
Stratocumulus type II (60−100%), $n=88$ and 384	0.25	1.18	−0.41	−0.67	0.46
Stratocumulus and cumulus type I (10 − 50%), $n=10$ and 35	−0.33	−0.42	0.35	1.71	0.85
Stratocumulus and cumulus type I (60 − 100%), $n=101$ and 179	−0.26	−1.06	−0.23	0.56	0.30
Stratocumulus and cumulus type II (10 − 50%), $n=15$ and 18	−0.22	2.30	0.66	0.22	−1.84
Stratocumulus and cumulus type II (60 − 100%), $n=84$ and 113	−0.02	0.58	−0.13	−0.52	−0.77

Note. n is the number of cases over which averaging was conducted. The first number for n corresponds to the surface and 300 mb levels, and the second to the 850, 700 and 500 mb levels.

The mean thermal advection in the absence of clouds does not exceed 0.05°/hr (about 0.6°/12 hrs), according to both Meteor and ESSA photographs. The difference lies in the sign of the advection. According to Meteor data the sign changes with height in the absence of clouds, and in the upper troposphere (at 300 mb) it becomes cold advection. According to data from ESSA, whose TV camera resolution is considerably lower, the advection is negative throughout the troposphere. This discrepancy is entirely admissible. A high-resolution camera allows one to pick out cases of complete absence of cloud cover, and these data confirm the opinion that, with small advection values of either sign, additional conditions are required for the appearance of clouds. On the other hand, a camera with much lower resolution views the region covered by scattered small cumulus clouds as if it were cloud-free. Such clouds often appear with cold air advection. This fact also determined the sign of the mean value of advection in the absence of clouds on the basis of ESSA pictures.

In the presence of stratiform clouds the deviation in individual advection values σ_A and ρ_A from the mean is revealed by their large scatter. Stratiform clouds can be observed not only with warm advection, but also with cold advection. However the mean values of advection are positive in all layers of the troposphere, and at some levels they reach about $2.5°/12$ hrs.

The mean values of thermal advection with cumulus clouds are considerable and have negative signs. Nevertheless cold advection decreases from the surface up to 700 mb and then increases again from the middle troposphere upward. With small amounts of cumulus there is a little warm advection near the ground which yields to cold advection. The mean value of cold advection does not exceed $0.5°/12$ hrs, whereas with considerable amounts of cloud in the upper half of the troposphere the mean value of cold advection is approximately $2.5°/12$ hrs.

Data on the distribution of advection with stratocumulus clouds are particularly interesting. It appears that with the same cloud structure as seen on TV pictures the vertical advection profiles may be exactly the opposite. Whereas a uniform cloud deck is only stratiform or only cumuliform and is associated with a uniform vertical distribution of advection, a more complex cloud cover, including at the same time cumulus and stratiform types, shows also a more complicated vertical thermal advection profile.

When we study computed advection values, two quite definite vertical advection profiles are discernible with the same distribution of signs over the cloud field in which stratocumulus clouds were observed.

In one case there was a positive change of temperature advection with height (type I), in the other a negative change (type II). In the first case warm advection increased, or cold advection decreased, or cold advection gave way to warm advection. In the second case cold advection increased, or warm advection decreased, or warm advection gave way to cold advection.

This particular feature in the distribution of advection in the presence of complex cloudiness is also distinguishable in the mean values of advection. Such a distribution of advection indicates the presence of multilayered clouds of various types at different tropospheric levels.

If one compares mean values of advection for clouds as determined by different satellites, it appears that the difference in the advection profiles stands out more sharply in Meteor cloud pictures. For clouds determined on the basis of ESSA pictures the vertical

profiles of values of $\overline{A}$ are much more smoothed. Here, the degree
of resolution of the TV camera is significant. The more detailed
Meteor pictures allow a stricter and more accurate classification
of clouds. However, the orders of values of advection and its vertical
profile for different cloud types as determined by Meteor and ESSA
photographs were entirely comparable and uniform.

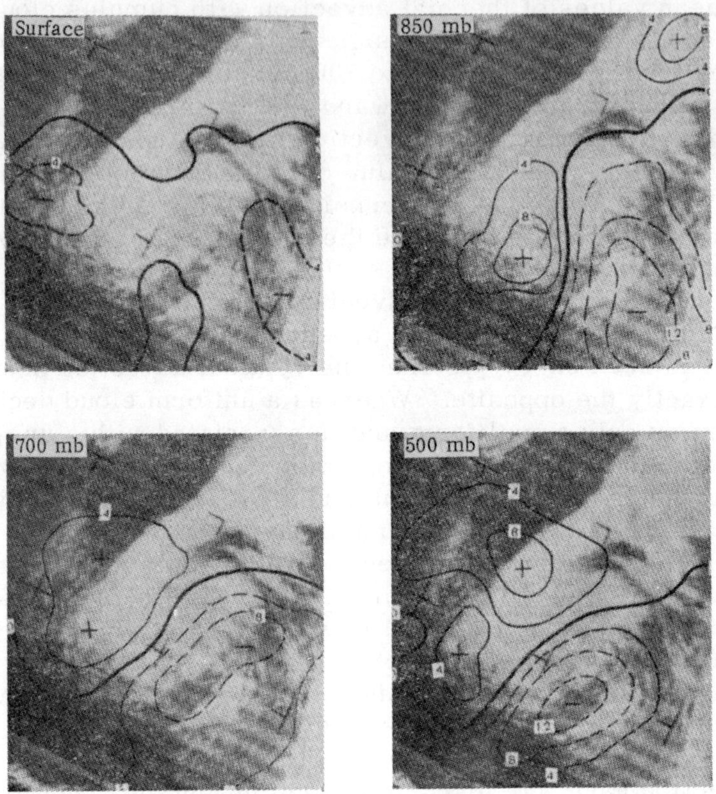

FIGURE 5.   Distribution of thermal advection (deg/ hr ·10) in a cloud
vortex on 15 May 1967.

As an example of the distribution of thermal advection in a cloud
vortex of an occluded depression we shall examine advection maps
of 15 May 1967 (Figure 5). The center of the depression is filled by
cold air, but the computed thermal advection shows that its thermal
asymmetry can be clearly discerned in the troposphere. Cold advection takes place in all parts of the depression only near the
earth's surface. From 850 mb upward two very sharp advection
centers can be seen, one positive and the other negative. In addition
there is a large advection gradient in the center of the depression.

The intensity of advection centers hardly changes with height in the lower half of the troposphere, but their axes are tilted in space. Below, the center of the cold pool extends from the parts of the cloud vortex with few clouds southward, while upward it becomes displaced nearer to the vortex center. The center of warm advection, in contrast, is displaced with height from the head of the cloud vortex along the main spiral, which consists of stratiform clouds.

The case of an occluded depression examined here is sufficiently typical to confirm that our assessment of the distribution of advection relative to the depression's cloud vortex is indeed representative.

Some of the work contained in this paper was prepared by candidates for higher degrees, L. M. Zhulimov, V. V. Fedorov and M. A. Bakiev, at the Hydrometeorological Center of the USSR under the supervision of the author.

The author is indebted to O. I. Mit'kin and G. K. Kriskevich for their technical help in the preparation of material.

UDC 551.509.311:551.515

CLOUD VORTICES IN THE MEDITERRANEAN AND SOME RELATED SYNOPTIC CONDITIONS

K. I. Stanchev, V. G. Sharov
(Bulgaria)

Cloud systems over the Mediterranean are examined for winter. Morphological features of cloud vortices are revealed and the relation between cloud vortices and synoptic conditions established. Some quantitative characteristics of cloud vortices are given.

Television photographs from artificial earth satellites give a clear picture of the areal distribution of cloudiness over regions commensurable with basic synoptic configurations. Morphological features, which remain undiscovered by discrete observations from the surface, can be discovered, so yielding some idea of the diversity of cloud systems.

One of the most typical morphological features of the cloud cover as observed from satellites is the spiral or vortex-like structure. Of special interest in this respect is the fact that these features are closely related to cyclonic pressure systems /1/. If this is so, one should be able to identify synoptic processes over the Mediterranean as "seasonal centers of cyclonic activity" by observing the morphological features of the cloud field, more specifically, cloud vortices over this region. This could be invaluable in the diagnosis of synoptic processes over the Mediterranean and immediate regions.

The value of satellite data for the Mediterranean is of particular importance owing to the sparsity of regular aerological information, both for the Mediterranean basin and for the north coast of Africa. In addition, one should remember that in the cold season snow cover can remain only over comparatively small parts of this region, and can therefore be easily discerned by its typical appearance.

This investigation presents some relationships between cloud vortices and related synoptic conditions over the Mediterranean.

At the same time we tried to demonstrate the potentialities of primary diagnostic, manifest and synoptically significant specific features in the cloud field. The case of a frontal spiral without a vortex band, or other features caused by the evolution of cloud vortices and related synoptic conditions are not treated here, since this will form the subject of a special investigation.

For the purpose of this research we chose the region 0 to 35°E, 30 to 47°N. Conditionally we call this the Mediterranean region, since it hardly differs from the actual areas in which southern depressions are born and develop.

CHARACTERISTICS OF INITIAL DATA

Initial data consisted of TV pictures of meteorological satellites Nimbus II and ESSA 2, 4, 6 and 8 over the Mediterranean region. Taking account of the seasonal activity of the "Mediterranean center" we examined the material for the cold season. For the periods 1 January to 31 March 1967, 1 October 1967 to 31 March 1968, and 1 October 1968 to 31 March 1969 we chose 60 cases when comparatively well developed vortex cloud structures were observed over the Mediterranean region. In addition two cases for April 1967 (Table 1) were investigated. The satellite data related to the period between 0830 and 1030 hrs. Therefore, when analyzing synoptic conditions we used surface observations for 0600 and 1200 GMT and aerological observations for 1200 GMT. Thus, the difference between satellite observations and the upper air observations did not exceed $3\frac{1}{2}$ hours. Nevertheless, though the surface observation of 0600 hrs was sometimes nearer to the satellite observations, it was decided (for uniformity of aerological analysis) to use mainly the 1200 GMT surface observations and the 0600 GMT observations as supplementary information.

GENERAL CHARACTER OF CLOUD PICTURES OVER THE MEDITERRANEAN REGION

The TV photographs showed that over the Mediterranean region no continuous cloud cover is observed and that it occupies at most 40 — 50% of the surface. This conclusion was drawn not only on the basis of the selected cases, but also because of the impression obtained in the process of choosing the basic data.

A dense cloud cover is typical for frontal zones. In these zones the most frequent cloud types according to surface observations are Sc, Ns and Cb. The width of the cloud band is usually 300 — 350 km, and very seldom exceeds 400 km. Rain areas are

nearly always associated with these bands.   Precipitation zones
are very difficult to trace over the Mediterranean because most of
it is sea area, and regular radiosonde reports are almost absent;
the same applies to the adjoining regions of North Africa.   This
makes it very difficult to relate rain areas to cloud systems.

A characteristic cloud system, quite often observed under certain
synoptic conditions over the Mediterranean, is shown in Figure 1.
This configuration (indicated by an arrow) bears a resemblance
to "drapes" or curtains and is distinguished from the cirrus type,
associated with a jet stream, in that it has much clearer outlines.
This stands out particularly in its southern part where the clouds
are broken up.   It suggests that in the examined case we can identify here a cloud system which differs from that described in /3/.*
An analysis of surface synoptic reports shows that these "drapes"
always consist of an upper layer of cirrus and a layer of middle
clouds, i. e., As and Ac.

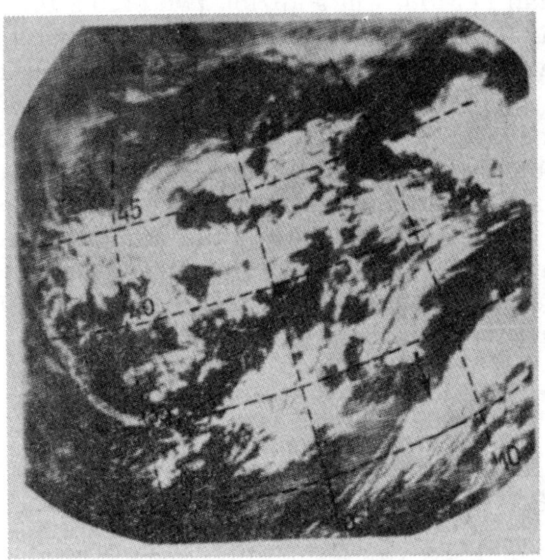

FIGURE 1. Cloudiness in the form of "drapes," associated with an upper-level frontal zone. 27 February
1968, 1054 GMT.

* [There is no reference /3/ in the Russian original.]

Usually, these "drapes" are seen over more southerly regions; their width is 400 — 500 km and their length 1200 — 1300 km. A comparison with radiosonde observations shows that this type of cloud system is associated with an upper-level frontal zone (UFZ), or more specifically with the front part of an upper trough. If we trace its longitudinal axis of symmetry the clouds are most often drawn out along the southwesterly flow in the immediate vicinity of the axis of the trough, or concentrated 100 — 150 km to the right of the axis of a frontal zone, i. e., on its warm side at the 500-mb level.

It is remarkable that the "drapes" which we deal with here are not connected with a cloud vortex, and in most cases the latter is not observed at all.

As expected, one of the general features which are typical for all cloud systems observed over the Mediterranean region is the fact that their thickness and their horizontal coverage of the ground diminish rapidly toward the southern part of the cloud system. There is also a difference in the cloud types: the northern parts are characterized by more stratified types while in the south cumulus forms prevail.

CLOUD VORTICES AND THEIR RELATED FEATURES

Morphological features of cloud vortices. Frontal cloud bands connected with cloud vortices, as observed over the Mediterranean, are either more weakly marked than those over more northerly latitudes, or altogether absent.

TABLE 1

Month	Gradation			Total
	1	2	3	
October	—	1	1	2
November	4	4	5	13
December	2	7	7	16
January	2	6	7	15
February	3	3	2	8
March	3	1	2	6
April	—	2	—	2
Total	14	24	24	62

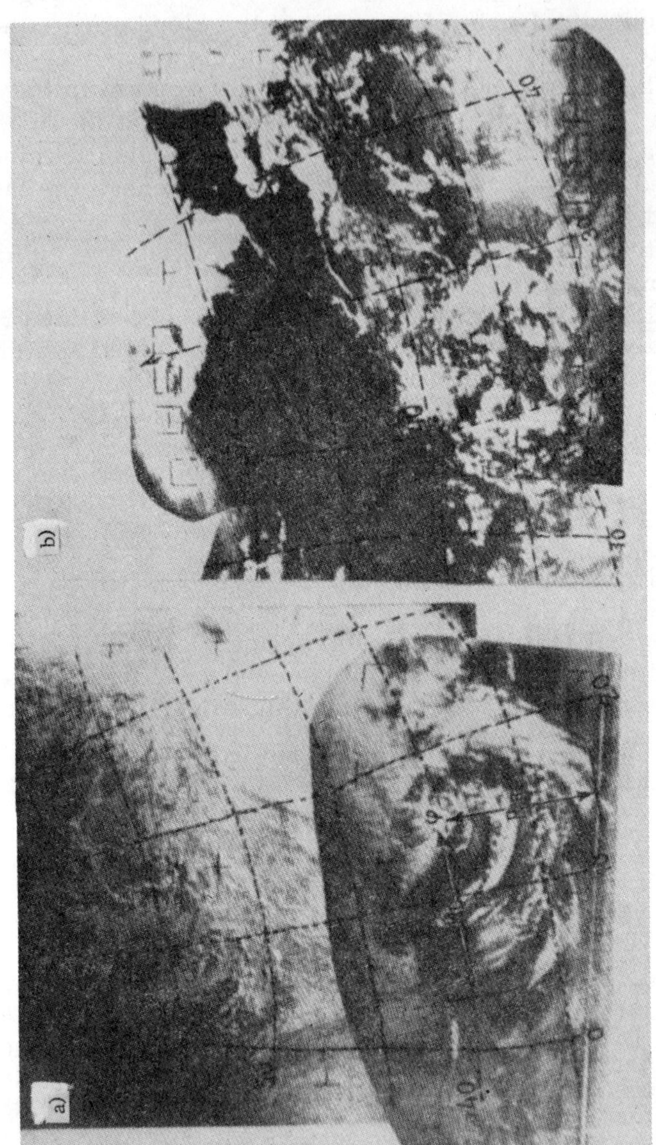

FIGURE 2. Types of cloud vortices: 1 — type I, 8 January 1968, 0922 GMT b — type II, 23 October 1968, 0937 GMT.

Table 1 gives the frequency of frontal belts associated with Mediterranean vortices in three gradations: 1 — well marked frontal belt, 2 — weak frontal belt, 3 — frontal belt absent.

An analysis of a great number of pictures showed that cloud vortices over the Mediterranean region consist as a rule of cumuliform clouds, a fact which, in our opinion, is explained by specific conditions of the atmospheric temperature stratification. This was confirmed also by surface observations which showed the presence of cumulus clouds. In the northern parts of the Mediterranean cloud vortices often consist of stratiform clouds.

The cases studied by us did not reveal, in addition to the main vortex, any secondary characteristic features, such as those described in /2/. However, in two cases cloud vortices were observed with dimensions typical of secondary cloud vortices, and in their vicinity the surface pressure field as well as the geopotential distribution in the middle troposphere showed weak troughlike characteristics. In one of these cases cyclogenesis took place, in the other it did not.

According to the morphology of cloud features in the Mediterranean cloud vortices can be divided into two types. Cloud vortices of type I (Figure 2a) are usually distinguished by the presence of frontal cloud spirals (in many cases weakly developed). Vortex belts in the central and rear parts are well marked, have a spiral form, and converge distinctly in the vortex center.

Cloud vortices of type II (Figure 2b) usually do not have frontal spirals. Vortex spirals are more eroded, and with increasing distance from the center they take on concentric circular forms. Research shows that these two types of cloud vortices are also notoriously different with regard to their respective synoptic conditions, on which we shall dwell further.

Spatial distribution of cloud vortices. During the investigation we tried to establish the spatial distribution of the observed cloud vortices. Figure 3 shows the frequency distribution of cloud vortex centers over the Mediterranean region. The frequency was computed for areas limited by 5° longitude and latitude lines. Considering that between 30 and 45° latitude the map scale hardly affects the accuracy employed for our investigation, these areas can be considered equal in size. The resulting frequency values were assigned to the center of the above areas. The isolines were drawn by linear interpolation. Figure 3 shows that the maximum occurrence of cloud vortices is displaced toward the eastern part of the region.

Doubtlessly this distribution reflects mainly the known relation between the cloud vortices and the state of development of depressions.

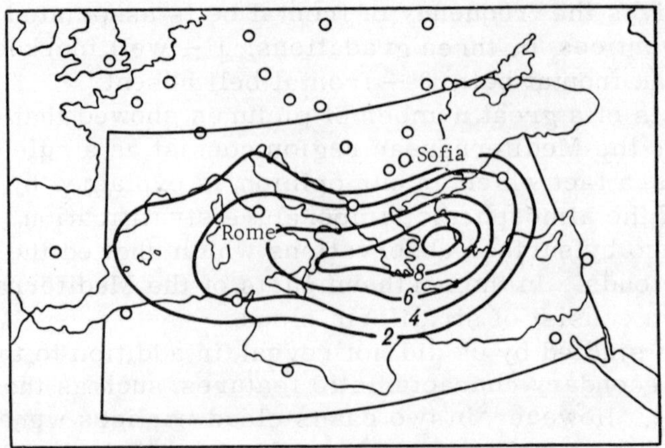

FIGURE 3.   Frequency distribution of cloud vortex centers over
the Mediterranean region.

Typical dimensions of cloud vortices.   It is well known that the
determination of the horizontal dimensions of cloud vortices is considerably hampered by their asymmetry, a property which is even
more marked in the Mediterranean by the nonuniform appearance
of frontal cloud spirals.   Bearing this in mind and for greater ob-
jectivity, we took as horizontal dimensions two radii: R_L along the
latitude lines and R_V along the meridians, respectively to the west
and south of the vortex center.   From Figure 2a it can be seen that
R_L and R_V are determined as the distance from the vortex center
to the corresponding tangent to the cloud mass toward the west and
south, respectively.   These, as can be seen, are not radii of one and
the same spiral.

Table 2 gives the frequency (No. of cases) of R_L and R_V of different lengths for 57 cases.   The maximum length for the meridional
radius is seen to be 600 km and for the zonal radius 700 km.   This
difference points to the asymmetry of cloud vortices which increases
with their dimensions.

TABLE 2

	Radius (km)										
	200	300	400	500	600	700	800	900	1000	1100	1200
R_L	0	1	3	6	9	11	10	7	5	3	2
R_V	1	2	7	11	12	11	7	3	2	1	0

The investigation of symmetry shows the frequency distribution of n values to be

$$\varepsilon = \frac{R_L - R_V}{R_L}.$$

For intervals of $\Delta\varepsilon = 0.1$ this yields:

ε	−0.25	−0.15	−0.05	0.05	0.15	0.25	0.35	0.45
n	1	1	0	24	12	9	7	3

The table shows that the vortices are drawn out along the west — east axis ($\varepsilon > 0$), and in only two case $\varepsilon < 0$. The mean of the examined values is $\varepsilon = 0.14$. Cases with $\varepsilon = 0$ were assigned to the inverval 0.0 to 0.1.

Cloud vortices and some related synoptic conditions. Owing to insufficient aerological and wind data over the Mediterranean region any definition of zones of wind convergence and divergence is very unreliable. Accordingly, the relation between cloud vortices and these zones was not investigated. Such a comparison was carried out for isallobaric centers, but no simple relation between them and cloud vortices could be found. The assumption that cloud vortices are related to negative pressure tendencies was not confirmed. In approximately 50% of cases positive pressure tendencies were found at the surface as well as at upper levels over regions with cloud vortices. Special attention was paid to research into relations between cloud vortices and the thermal and pressure field.

For this purpose we used synoptic surface maps and maps of the actual topography of 700 mb (AT_{700}) and maps of the relative topography of 1000/500 mb (RT_{1000}^{500}). With regard to the synoptic conditions of the earth's surface, the conclusions drawn in /1/ were fully supported, and therefore we shall not dwell on these further.

The relation between cloud vortices and the mean temperature advection in the 1000/500 mb layer was investigated. It was shown that the vortex structure of the cloud field was more clearly marked, the more prominent the geostrophic cold advection.

At the same time the center of the cloud vortex almost coincided with the cold core as marked either in the relative topography field or only in the thermal field in the same layer. A number of cases with cloud vortices were observed in which for some reason no definite cold core was observed. In these cases the center of the cloud vortex was usually found in the immediate vicinity of the axis of the thermal trough at a distance of 150 − 200 km to the southwest of the pressure center at 700 mb.

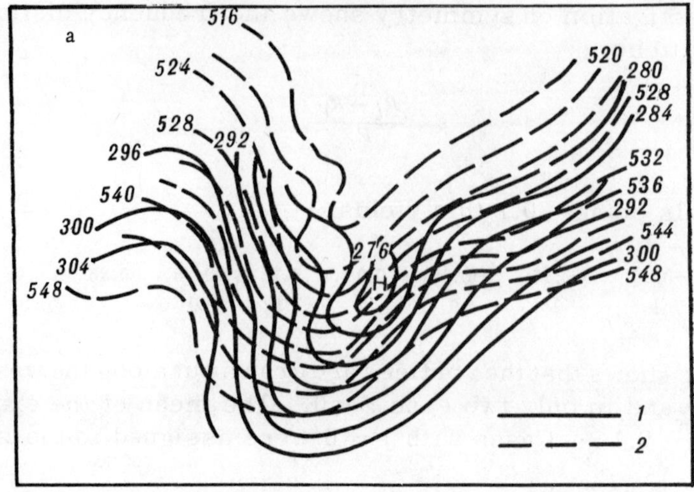

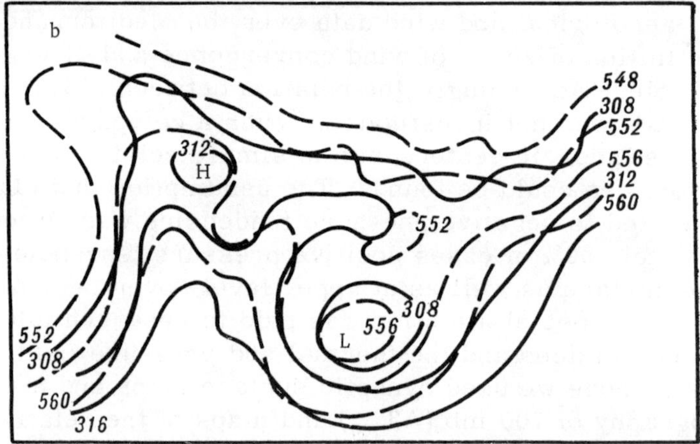

FIGURE 4.   Thermal and pressure field typical for cloud vortices of different types:

a — type I, 8 January 1968, 1200 GMT; b — type II, 23 October 1968, 1200 GMT; 1 — isolines of AT_{700}; 2 — isolines of RT_{1000}^{500}.

For cloud vortices of type I (such as that shown in Figure 2a) the typical thermal and pressure field has the form shown in Figure 4a.   In this case the cloud bands are orientated in such a way that the angle between the band and the isolines of RT_{1000}^{500} is only half that between the cloud band and the isolines of AT_{700}.

In addition, the area occupied by the cloud vortex of type I is limited by the curve describing zero advection in the layer 1000/500 mb.   This curve nearly always lies outside the rounded vortex.   In this case it is evident that for the observation of a well formed and nearly symmetrical vortex the curve of zero advection must be closed, if there is to be transport of the cold pocket.

Cloud vortices of type II (Figure 2b) and their related thermal and pressure fields as shown in Figure 4b are typical for the stage following the maximum development and the beginning of filling of the depression.   Then, the cold core coincides with the pressure center and as a consequence the thermal advection becomes weak or nonexistent.   In this case the curve of zero advection is not clearly defined and the dimensions of the cloud vortex depend on the dimensions of the depression proper.   These vortices are usually symmetrical, i. e., $\varepsilon = 0$ and their radii seldom exceed 500 km.

CONCLUSIONS

Our conclusions do not contradict the results obtained in /1/, which deals with the position of cloud vortices in relation to depression centers.   This applies even more to the diagrams given in that work showing the position of cyclone centers in relation to the cloud vortices.   These diagrams are very similar to diagrams of the position of thermal troughs in relation to pressure centers.   The establishment of this fact on a theoretical basis could be of fundamental significance.   It would mean that baroclinic effects are a basic occurrence during vortex formation and not "vorticity advection," a result which contradicts the widely held opinion about the quasibarotropic behavior of the atmosphere in synoptic processes.   Of no less theoretical importance is the appearance of asymmetry in the cloud vortices.   Until now one could raise objections against the wholesale application of this conclusion to the vortex circulation, since it is not known to what degree this asymmetry may be explained by the effect of erosion of clouds, an effect which is not directly connected with the horizontal circulation.   This fact certainly cannot be disregarded and, in our opinion, is related to the horizontal structure of depressions.

BIBLIOGRAPHY

1. Leonov, N. G. et al.   Sinopticheskie usloviya sushchestvovaniya oblachnykh vikhrei (Synoptic Conditions during the Appearance of Cloud Vortices). — Trudy Gidromettsentra SSSR, No. 11. 1967.
2. Popova, T. P. and A. M. Tsar'kova.   "Vtorichnyi" oblachnyi vikhr' ("Secondary" Cloud Vortices). — Meteorologiya i Gidrologiya, No. 9. 1967.

UDC 551.576

ANALYSIS OF PRESSURE AND CLOUD ANOMALY FIELDS

N. A. Chuzavkova

This paper treats the correspondence between fields of mean monthly surface pressures and geopotential height at different atmospheric levels and fields of total cloudiness. Also examined is the relationship between fields of deviation from climatological means of these values for July and October 1965, and for January and April 1966. Analogy criteria are given for the fields of mean monthly pressure and geopotential for 850, 700 and 500 mb and the total cloudiness.

Anomaly profiles for the indicated meteorological elements are analyzed for 60°N. The degree of similarity between fields of pressure anomalies and cloud anomalies are determined according to analogy criteria calculated for the above months. A qualitative comparison is made between anomaly maps of the geopotential at 700 mb and the total cloudiness for January 1966.

With meteorological satellites presently viewing the cloud cover over practically the entire globe, the question naturally arises as to how closely the cloud distribution corresponds to a number of other meteorological elements.

It is well known, for instance, that as a rule depressions are associated with large fields of complex cloudiness, while their center is often connected with large cloud vortices. In contrast, weather in anticyclones is usually connected with cloudless skies or with small cloud amounts. As regards large-scale atmospheric processes affecting vast regions, cases arise in which the cloudiness does not characterize in any definite way the field of pressure or absolute topography, because no simple relationship exists between them. However, to solve the problem of reconstructing the total pressure field on the basis of cloud distribution, we require the relationship between cloud distribution and the pressure field over large regions or even over a hemisphere. In a number of investigations the global seasonal average cloud amount is determined in the form of maps of the mean global cloud cover established by satellite data. Arking /4/ examined the latitudinal cloud-cover distribution from Tiros III photographs for summer 1961. Clapp /5/ investigated the global cloud cover using Tiros nephanalyses during 1962 — 63. Clapp's cloud data served to establish world maps and latitudinal profiles of the average cloud amount for the four seasons

from March 1962 to February 1963. He also discovered that anomalies of the cloud cover are associated with anomalies of the general circulation.

The object of the above-cited works was to establish a relation between mean monthly pressure fields at different atmospheric levels and the distribution of total cloud amount. With this aim tables of mean monthly pressure, absolute topography at 850, 700 and 500 mb, and total cloud amount were drawn up. The cloud cover was obtained from meteorological satellites over the northern hemisphere for July and October 1965 and for January and April 1966. These data were taken from a grid with steps of latitude $\Delta\varphi = 5°$ and of longitude $\Delta\lambda = 10°$.

The degree of similarity between the above meteorological fields was determined by computing the criteria of gradient analogies ρ_φ and ρ_λ where ρ_φ and ρ_λ are the values of criteria ρ (Table 1) calculated for the gradients with respect to φ and λ, respectively (φ is latitude and λ longitude). These quantities for October 1965, say, were as follows: $\rho_\varphi = 0.18$, $\rho_\lambda = 0.16$ for the earth's surface. For isobaric levels 850, 700 and 500 mb the values of ρ_φ and ρ_λ were, respectively, 0.24 and 0.11, 0.24 and 0.16, 0.17 and 0.12. On the basis of these values some analogies were observed between the investigated fields, but they were not pronounced. The question then arises: if the connection between fields of mean monthly pressure and cloudiness is not very pronounced, what is the character of the relationship between the anomaly fields of the examined meteorological elements? To provide an answer we used mean monthly values of pressure and cloudiness and their climatic values to compute deviations from normal pressure /3/, absolute topography /1/ and total cloud amount /2/ for the above-mentioned months. The deviations were computed in the usual way: if $x_{\text{mean month}}$ is the mean monthly value of any meteorological element and x_{clim} its normal value, the deviation is

$$\Delta = x_{\text{mean month}} - x_{\text{clim}}.$$

Values of the deviation are entered into the corresponding tables. These data were used to construct latitudinal profiles of the distribution of anomalies of pressure, geopotential height of the 850, 700 and 500 mb surfaces, and of the cloudiness for July and October 1965 and for January and April 1966, between 60 and 40°N latitude. The analogy of these profiles of the above-indicated anomalies led to the conclusions that, in general, a pressure rise is associated with decreasing cloudiness and pressure drops with increasing cloudiness. As an example we present the profile of the

anomaly of pressure, of the geopotential at the 850, 700 and 500 mb
surfaces, and the cloudiness for January 1966 along latitude 60°N
(Figure 1). A comparison of the curves shows that the positive
pressure anomaly in the region 0 to 70°E and 130 to 40°W along the
60° latitude line corresponds to a negative cloud anomaly over these
regions, while the negative pressure anomaly over the regions
between 70 and 130°E coincides with the positive cloud anomaly over
these same regions. Naturally, there is no complete correlation
between curves a and b. This is particularly pronounced in the
region 140°E to 130°W along 60°N (Pacific Ocean). Here the cloud
anomaly is near zero, while the anomaly of pressure and geopotential
height is negative. Possibly, in this region the analysis of the pressure and geopotential height and cloudiness was not sufficiently accurate owing to the sparcity of data pertaining to this region.

The tables of anomalies of pressure, geopotential height and
cloudiness were used to construct anomaly maps for the northern
hemisphere for July and October 1965, and January and April 1966;
they provide visual comparison of the fields. Their analysis shows
that over Europe, Greenland and the Atlantic Ocean there is, on the
whole, conformity between the above fields, i. e., negative (positive)
pressure and geopotential anomalies correspond generally to positive (negative) anomalies in cloudiness.

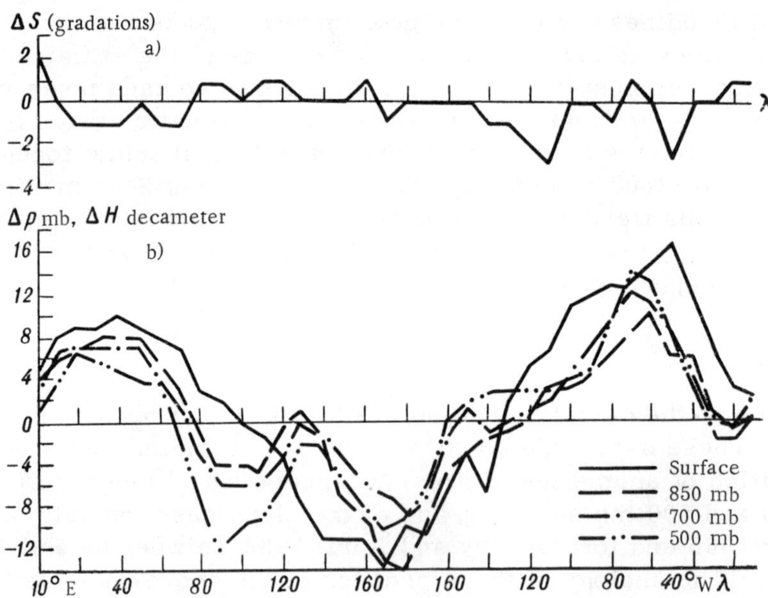

FIGURE 1. Anomaly profile of the mean monthly cloudiness along 60°N (a) and
the anomaly profiles of mean monthly pressure, geopotential height at 850, 700
and 500 mb along 60°N (b), for January 1966.

However, in a number of cases this correspondence is observed
only at certain levels. For instance, comparison of the anomaly
maps for January 1966 indicates the following facts. In the Scandinavian peninsula and north European USSR positive pressure anomalies extend to the 700-mb level and coincide with negative anomalies of total cloudiness. Over the Balkan region negative pressure anomalies extend up to 500 mb and coincide well with positive
anomalies of the total cloud amount. A similar picture of fairly
good correspondence between the above-indicated meteorological
elements is observed in several other regions, e. g., at the 850-mb
level over the North Atlantic Ocean near North America.

Nevertheless there are also regions where negative anomalies
of pressure and geopotential height correspond to negative anomalies
of total cloudiness. Such an example is Siberia up to the 850-mb
level. There were also instances in which considerable pressure
and geopotential anomalies corresponded to zero anomalies of
cloudiness.

A more detailed comparison of the anomaly map of mean monthly
cloudiness (Figure 2) with that of mean monthly geopotential heights
for 700 mb (Figure 3) shows that in a number of regions of the northern hemisphere, such as the Appenine and Balkan peninsulas, the
Black Sea, Siberia from Lake Balkhash to Lake Baikal and the Gulf
of Mexico, a positive anomaly of the mean monthly cloudiness corresponded to a negative anomaly of the mean monthly geopotential
height, while in the region of Scandinavia, Greenland and Central
Asia negative anomalies of mean monthly cloudiness corresponded to
positive anomalies of the mean monthly geopotential height. The
Atlantic Ocean stands out particularly. There, considerable negative anomalies of the geopotential coincide with practically zero
cloud anomalies.

TABLE 1. Values of analogy criteria ρ

Period	Surface, mb			
	Earth	850	700	500
July 1965	0.27	0.38	0.35	0.19
October 1965	0.52	0.49	0.40	0.32
January 1966	0.40	0.34	0.26	0.10
April 1966	0.49	0.37	0.41	0.36
Average	0.40	0.40	0.33	0.24

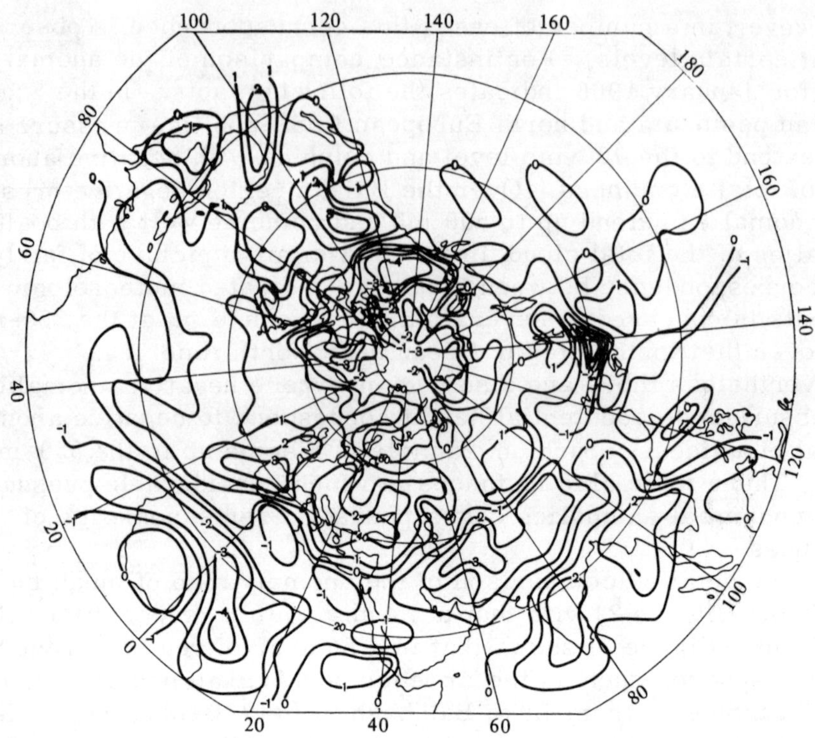

FIGURE 2.   Anomaly map for mean monthly cloud amounts.   January 1966.

From the above qualitative analysis we may conclude that in the majority of examined cases there is a correlation between pressure and cloudiness anomalies of opposite sign.

For objective assessments of the degree of similarity in the anomalies of the examined meteorological elements we computed analogy criteria according to the formula $\rho = \dfrac{n_+ - n_-}{n_+}$, where n_+ is the number of grid points at which the anomaly of pressure (or of geopotential height) and the anomaly of cloudiness have opposite signs, and n_- the number of grid points at which the signs of the anomalies are identical.   The case when one of the compared values is zero was disregarded.   Values of ρ are listed in Table 1.

The data analyzed in Table 1 led to the conclusion that the estimated values obtained for the 500-mb level were the lowest.   The values given in the table further imply that for transition months (October, April) the values of ρ were higher than for January and July.   The following fact is also noteworthy: if we designate the values of ρ for the surface, 850, 700 and 500 mb levels by ρ_0, ρ_1, ρ_2, ρ_3, respectively, it is seen that $\rho_0 > \rho_1 > \rho_2 > \rho_3$.   This means that the connection between the distribution of anomalies of cloudiness and pressure at different levels weakens with height.

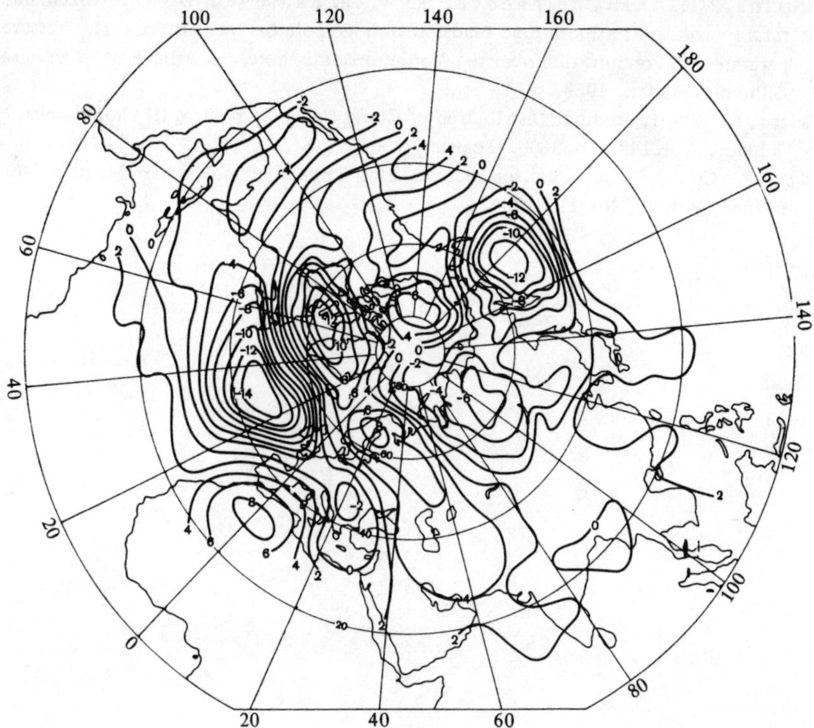

FIGURE 3.   Anomaly map for mean monthly geopotential heights for 700 mb. January 1966.

Analysis of the data presented in Table 1 also showed that positive anomalies of cloudiness coincide in approximately 70% of cases with negative anomalies of surface pressure and of the geopotential at the 850-mb level.   For higher pressure levels the percentage is somewhat less.

It follows that in some problems of satellite meteorology it is advisable to analyze the fields of cloudiness and pressure (or geopotential height) anomalies instead of simply studying the cloud and pressure fields.

BIBLIOGRAPHY

1. Aeroklimaticheskii atlas severnogo polushariya (Aeroclimatological Atlas of the Northern Hemisphere). Vol. 1. Leningrad, Gidrometeoizdat. 1961.

2. Musaelyan, Sh. A. Klimaticheskie dannye ob oblachnom pokrove Zemli i perspektivy ikh ispolzovaniya pri reshenii zadach sputnikovoi meteorologii (Climatological Data for the Earth's Cloud Cover and Prospects for their Application to Problems of Meteorological Satellites). — Trudy Gidromettsentra SSSR, No. 30. 1968.

3. Sokhrina, R. F., O. M. Chelpanova, and V. Ya. Sharova. Davlenie vozdukha, temperatura vozdukha i atmosfernye osadki severnogo polushariya (Atmospheric Pressure, Temperature and Precipitation over the Northern Hemisphere). — Atlas Kart. Leningrad, Gidrometeoizdat. 1959.

4. Arking, A. The Latitudinal Distribution of Cloud Cover from Tiros III Photographs. — Science, Vol. 143, No. 3606. February 1964.

5. Clapp, P. Global Cloud Cover for Seasons Using Tiros Nephanalyses. — Monthly Weather Review, Vol. 92, No. 11. 1964.

UDC 551.576:551.558.2

POSSIBILITY OF COMPUTING THE CLOUD FIELD ON THE BASIS OF VERTICAL VELOCITIES

Z. N. Fatkhullaeva

A model is proposed for calculating the large-scale field of cloudiness from data on ordered vertical motion. Results are given of computing maps of total cloud amount as a diagnostic procedure. Results of computations are compared with actual data.

A method for computing cloud distribution on the basis of geostrophic vorticity is given in /2/. This method was used to establish a method of harmonic correlation /1/. The main point of the latter method is that data on cloudiness and geostrophic vorticity are arranged in trigonometric series for each line of the used grid for every time step. Regression equations are then constructed for each harmonic and for each line. The resulting system of equations was used to set up a system of standard equations by the method of least squares, whence we also determine the unknown regression coefficients.

In the present paper we investigate the possibility of computing the total cloud amount from data on vertical velocities. The computation model is described below.

In a rectangular coordinate system (the X-axis points east and the Y-axis north) we consider the area covered by a grid of 20×16 points with $h = 300$ km steps. In this area we examine a selection of data for the total cloud amount and vertical velocity at the mean layer of the atmosphere for a number of consecutive periods (time steps).

Let us designate the functions describing the field of total cloud amount and vertical velocities for the ν-th time step and along the j-th line ($j = 1, 2, 3 \ldots 16$), respectively, by $S_j^{(\nu)}(x_i)$ and $W_j^{(\nu)}(x_i)$ and let us examine the following expansions:

$$\left.\begin{aligned}
S_j^{(\nu)}(x_i) &= \frac{s_{j0}^{(\nu)}}{2} + \sum_m s_{jm}^{(\nu)} \cos \frac{m\pi x_i}{19h} \\[2mm]
W_j^{(\nu)}(x_i) &= \frac{w_{j0}^{(\nu)}}{2} + \sum_m w_{jm}^{(\nu)} \cos \frac{m\pi x_i}{19h}
\end{aligned}\right\} \qquad (1)$$

where $i = 1, 2, 3 \ldots 20$, $s^{(v)}_{jm}$ and $w^{(v)}_{jm}$ are Fourier coefficients for functions $S^{(v)}_j(x_i)$ and $W^{(v)}_j(x_i)$, and m is the wave number.

As in the above-cited papers we shall propose that the following linear relationship exists between the Fourier coefficients for cloudiness $s^{(v)}_{jm}$ and vertical velocities $w^{(v)}_{jm}$:

$$s^{(v)}_{jm} = \alpha_{jm} w^{(v)}_{jm} + \beta_{jm}, \tag{2}$$

where α_{jm} and β_{jm} are regression coefficients determined by the selected data.

Let us assume that the selection investigated here consists of k time steps ($v = 1, 2, 3, \ldots k$). Equation (2) for all time steps entering the selection assumes the following form:

$$\left.\begin{aligned}
w^{(1)}_{jm} \alpha_{jm} + \beta_{jm} &= s^{(1)}_{jm} \\
w^{(2)}_{jm} \alpha_{jm} + \beta_{jm} &= s^{(2)}_{jm} \\
\cdots \cdots \cdots \\
w^{(k)}_{jm} \alpha_{jm} + \beta_{jm} &= s^{(k)}_{jm}
\end{aligned}\right\} \tag{3}$$

From this system a set of standard equations can be constructed by the method of least squares, and hence the regression coefficients are easily found:

$$\left.\begin{aligned}
\alpha_{jm} &= \frac{\sum_{v=1}^{k} \left(w^{(v)}_{jm} - \overline{w}_{jm}\right)\left(s^{(v)}_{jm} - \overline{s}_{jm}\right)}{\sum_{v=1}^{k} \left(w^{(v)}_{jm} - \overline{w}_{jm}\right)^2} \\
\beta_{jm} &= \overline{s}_{jm} - \alpha_{jm} \overline{w}_{jm}
\end{aligned}\right\} \tag{4}$$

where

$$\overline{s}_{jm} = \frac{1}{v} \sum_{v=1}^{k} s^{(v)}_{jm},$$

$$\overline{w}_{jm} = \frac{1}{v} \sum_{v=1}^{k} w^{(v)}_{jm}. \tag{5}$$

According to formulas (3), (4) and (5) we compute the unknown coefficients of regression on the basis of the Fourier coefficients of the initial data on cloudiness and vertical velocities for the k previous time steps. Then, the resulting regression coefficients and the field of vertical velocities for the $(k+1)$ time step are used to compute, according to formulas (2) and (1), the unknown field of cloudiness for the $(k+1)$ time step.

The above model was employed in computing the field of cloudiness on the basis of data for the vertical velocity at the 700-mb
level.   Maps of the total cloud amount were constructed for the investigated areas from 14 to 31 May 1965 using TIROS IX data.   For
the same period data on vertical velocities were prepared for the
700-mb level at 1500 hrs, calculated daily in line with the program
and method of S. L. Belousov /3/.   These data were prepared for
18 observation times.

The initial data from which regression coefficients α_{jm} and β_{jm}
were computed were chosen as the data for the five previous successive time intervals.   The computed regression coefficients were
used in the calculation of the cloudiness field for the six following
intervals.   Thus, for the computation of the cloudiness field on
19 May 1965 the regression coefficients were calculated according
to data for the total cloud amount and vertical velocities on 14, 15,
16, 17 and 18 May.

In this way 13 maps were constructed of the total cloud amount
in gradations.   Each map contained data for $20 \times 16 = 320$ grid
points.   For each map, for 19 to 31 May, the following mean absolute
errors of the cloud field were computed:

Number	19	20	21	22	23	24	25	26	27	28	29	30	31
$\lvert \bar{\delta} \rvert$	2.8	3.2	3.0	2.6	3.0	3.0	3.3	3.0	2.8	2.9	3.3	2.8	2.3

These data show that the mean absolute error fluctuates between
2.3 and 3.3 gradations and the mean error for all examples was
2.9 to 3.6 gradations, while the mean error for the 13 constructed
maps was 3.2 gradations.   It follows that the use of data on vertical
velocities yielded somewhat improved results.

We now examine one of the computed maps.   Figure 1 presents
an actual map of total cloud amount, constructed on the basis of
nephanalysis of data received by TIROS IX on 28 May 1965.   In
areas where satellite data were missing these were supplied by
cloud amounts taken from synoptic maps.   Figure 2 shows a computed map of total cloud amount for 1500 hrs on 28 May 1965.   The
regression coefficients for this example were determined by data
chosen for 23 through 27 May 1965.   The cloud field was computed
by data on vertical velocities for 28 May.   Comparison of the maps
shows that, along general lines, they correspond.   Thus, one observes on Figure 1 predominantly cloudy weather with considerable
and complex cloudiness over central Europe.   Over the Baltic Sea
and the northern Black Sea a zone of few clouds extends.   All these
features are also reflected on the computed map of cloudiness in
Figure 2.

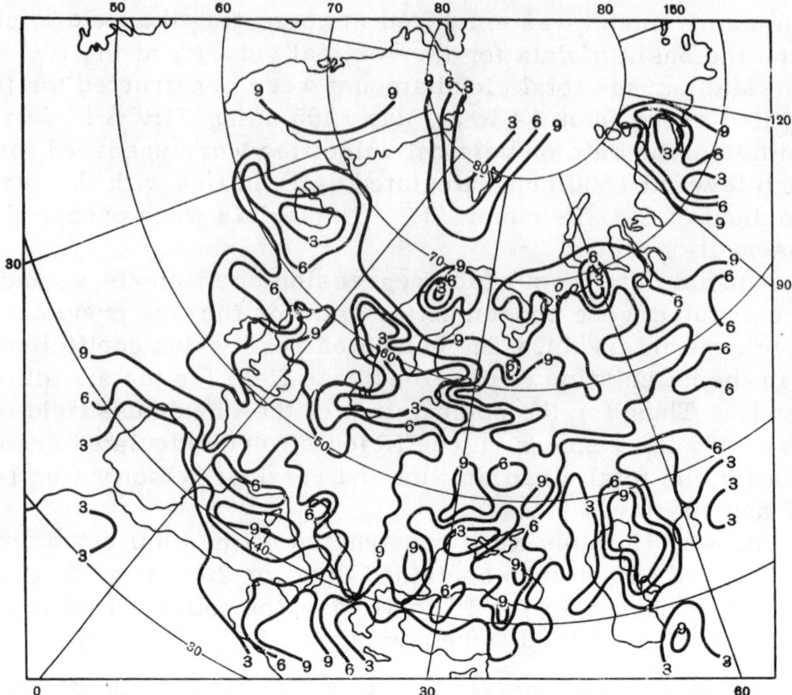

FIGURE 1. Actual map of total cloud amount at 1500 hrs on the basis of TIROS IX data for 28 May 1965.

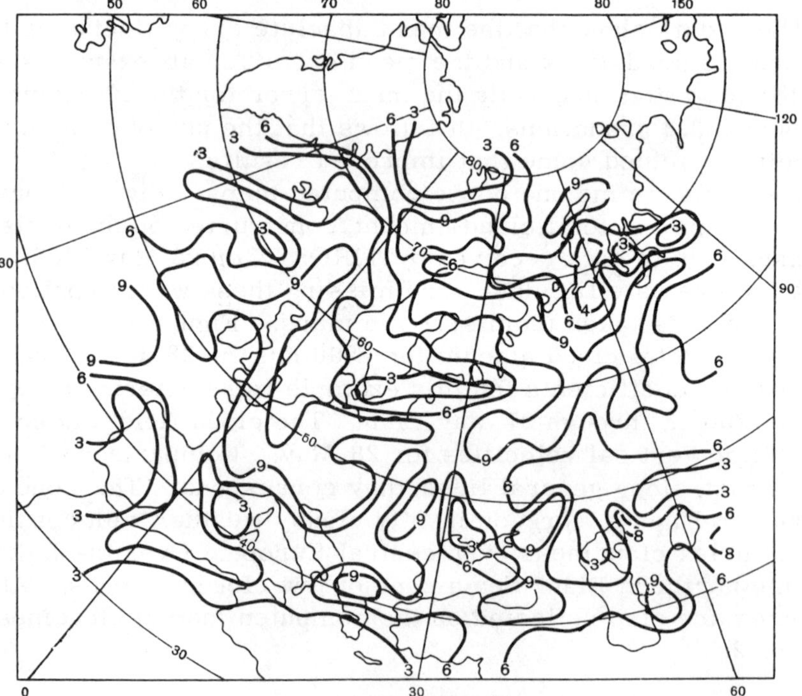

FIGURE 2. Computed map of total cloud amount at 1500 hrs on 28 May 1965.

Comparison of Figures 1 and 2 also indicates that there are regions in which actual and computed data do not correspond (northern Scandinavia and elsewhere).

The proposed model can serve for prognostic purposes if, instead of diagnostic vertical velocities for the $(k+1)$ time step, we use prognostic vertical velocities.

BIBLIOGRAPHY

1. Musaelyan, Sh. A. Nekotorye aspekty interpretatsii i ispol'zovaniya dannykh ob oblachnosti, poluchaemykh s meteorologicheskikh sputnikov (Some Aspects of Interpretation and Use of Cloud Data Received by Meteorological Satellites). — DAN SSSR, Vol. 163, No. 5. 1965.
2. Musaelyan, Sh. A. and Z. N. Fatkhullaeva. Ob odnom sposobe ispol'zovaniya dannykh ob oblachnosti, poluchaemykh s meteorologicheskikh sputnikov (A Method for Using Cloud Data Received by Meteorological Satellites). — Trudy Gidromettsentra SSSR, No. 11. 1967.
3. Rukovodstvo po kratkosrochnomu prognozu pogody, chast' I (Guide to Short-Range Weather Forecasting, Part I). Leningrad, Gidrometeoizdat. 1964.

UDC 551.571.7

METHOD FOR ANALYZING THE MOISTURE FIELD BY THE USE OF SATELLITE CLOUD DATA

N. A. Chuzavkova

This paper deals with the possibility of reducing mean absolute errors in computing the moisture field on the basis of satellite cloud data in data-sparse regions, by using actual data from a wide-mesh observation network. Agreement between the computed moisture fields and actual data is attained by polynomial interpolation of calculation errors. Values for mean absolute errors are given which are calculated from data on clouds and the corresponding moisture fields.

One of the main uses of satellite cloud data is in the reconstruction of fields of meteorological elements over regions lacking in observation stations. It has been shown /2/ that data on cloudiness and moisture for the n preceding consecutive observation times and cloud data for the $(n+1)$ observation period can be used to construct the moisture field for this time. The problem was solved for the northern hemisphere, but the results presented in /2/ have shown that calculated moisture fields are much more smoothed than the actual field. The mean absolute errors of calculation were considerable. As representative parameter for moisture we chose the dew point.

In order to reduce the mean absolute error of moisture fields calculated according to data on cloudiness in data-sparse regions, we compared these fields with actual data from a sparse station network. For this comparison we used the method worked out by Musaelyan and Aprausheva /1/. This method uses a polynomial interpolation of calculation errors introduced by the use of actual data from data-sparse regions. Since, as mentioned above, the dew-point distribution was computed for the northern hemisphere $(80-30°N)$ we can determine for every point in a data-sparse region the dew point computed on the basis of cloud data. Let us express these values by τ_s. In addition, we have for the same points the known actual values for the dew point τ_a for every observation period. Consequently, we can find accurate values of the absolute error of calculation for each station of the thin network: $\delta = \tau_a - \tau_s$. Another problem consists in determining the value of these errors at the points of a geographical grid covering this region. Then,

knowing the error δ and the values of the dew point τ_s calculated from cloud data at the points of the geographical grid, one can obtain the values for the dew point τ taking account of the actual data:
$\tau = \tau_s + \delta$.

We now describe in detail the method of determining the absolute error of calculation at the points of the geographical grid. At all points of our data-sparse grid we form triangles such that the above-mentioned geographical grid points fall within (or on the borders of) some triangle. In addition, we regard the examined areas as flat (on the map) in the Cartesian system of coordinates (x, y). Then, each geographical grid point and each station is defined by coordinates (x, y). A further description of the method will be provided by the example of one triangle with vertices ABC. Suppose a geographical grid point D falls within this triangle. The absolute computation errors of the dew points for the vertices can be expressed in the form

$$\delta^{(A)} = \tau_a^{(A)} - \tau_s^{(A)},$$
$$\delta^{(B)} = \tau_a^{(B)} - \tau_s^{(B)},$$
$$\delta^{(C)} = \tau_a^{(C)} - \tau_s^{(C)}. \tag{1}$$

Comparison of the computed with the actual field of τ_s shows that the variation in errors δ is smooth. One can therefore assume that inside triangle ABC the value of δ changes linearly, i. e.,

$$\delta = ax + by + c, \tag{2}$$

where a, b, c are some coefficients.

Equation (2) expressed for each vortex of triangle ABC is

$$\delta^{(A)}(x, y) = ax_A + by_A + c,$$
$$\delta^{(B)}(x, y) = ax_B + by_B + c,$$
$$\delta^{(C)}(x, y) = ax_C + by_C + c. \tag{3}$$

The values of a, b and c can be obtained from system (3). The absolute error $\delta^{(D)}$ at point D of the geographical grid point lying within triangle ABC is

$$\delta^{(D)}(x, y) = ax_D + by_D + c. \tag{4}$$

The relationship between quantities $\tau_s^{(D)}$ and $\delta^{(D)}$ and the dew point at D is

$$\tau^{(D)} = \tau_s^{(D)} + \delta^{(D)}(x, y). \tag{5}$$

Similar corrections apply to all geographical grid points in the region under consideration.

The model computation can be conducted over a region with sufficient meteorological data, within which a fictitious data-sparse network can be established. The data of this latter network agree with reduced cloudiness data and can be compared with the actual dew-point field. This gives a possibility of estimating to what degree computations with an actual data-sparse network permit moisture values to be calculated from cloudiness data in the region under consideration. Such a modeling procedure was conducted over Europe between $85-35°$N and $10°$W $-55°$E for a transference of data pertaining to a selected little-known region of the North Atlantic.

These computations showed that if the mean absolute error of the dew point, based on cloud data, was $4.2°$, after accounting for the actual data for the dew point in a region with a sparse network the error decreased to $2.0°$.

A trial of this method was made over a real data-sparse region (North Atlantic). Three examples were computed: one for the 850-mb level and two for the 500-mb level. For the 850-mb level the initial data were taken at 0300 hrs and for the 500-mb level at 1500 hrs. The computed results can be judged from the values for the mean absolute errors given in Table 1.

TABLE 1. Mean absolute computation errors of the dew point based on cloud data and taking into account data from a sparse observation network

Data and observation times	$\lvert \overline{\delta^{\circ}_{s}} \rvert$	$\lvert \overline{\delta^{\circ}} \rvert$
21 July 1965, 0300 hrs	2.6	1.7
12 July 1965, 1500 hrs	4.5	2.8
21 July 1965, 1500 hrs	4.3	2.3

Note. Here $\lvert \overline{\delta_s} \rvert$ is the mean absolute computation error of the dew point based on cloud distribution; $\lvert \overline{\delta} \rvert$ is the mean absolute computation error, taking into account actual data from a sparse observation network.

Analysis of data in this table shows that the mean absolute computation errors, taking into account the actual dew point data from a region with a sparse observation network, are smaller than the computation errors according to satellite cloud data. It is noteworthy that for the 850-mb level the errors are smaller than for the 500-mb level.

In this way one can assert that the use of actual data from data-sparse regions decreases the computation error and allows one to improve the quality of the dew point field, computed on the basis of cloud data, thus making it more similar to the actual field.   This improves analysis of the dew point field, and compensates to a certain extent for insufficient information.

BIBLIOGRAPHY

1. Musaelyan, Sh. A. and N. N. Aprausheva. O sovmestnom ispol'zovanii dannykh nablyudenii so sputnikov i redkoi seti aerologicheskikh stantsii dlya vosstanovleniya polei meteorologicheskikh elementov (On the Concurrent Use of Satellite Observations, and Observations from a Sparse Aerological Network in the Reconstruction of Fields of Meteorological Elements). — Trudy Gidromettsentra SSSR, No. 30. 1968.
2. Chuzavkova, N. A. O vozmozhnosti rascheta polya vlazhnosti po raspredeleniyu oblachnosti (On the Possibility of Computing the Moisture Field from the Cloudiness Distribution). — Trudy Gidromettsentra SSSR, No. 11. 1967.

UDC 551.571.7

USE OF SATELLITE CLOUD DATA IN MOISTURE FIELD ANALYSIS

V. V. Ozerkina

The possibility is examined of using satellite cloud data in the analysis of the dew point deficit field by the method of optimum interpolation. The statistical structure of this field is investigated for the 850-mb level in its relation to the cloud field. Normalized spatial autocorrelation functions and cross-correlation functions are calculated for the dew point deficit for two types of cloud cover.

Analysis of the moisture field is very difficult, especially in regions with widely spaced meteorological stations, since moisture is very variable in space. The scale of this variability can be indirectly assessed by satellite cloud observations and is closely connected with the moisture field. A particularly good correlation exists between cloudiness and relative humidity /1, 8 —10/.

In this paper we investigated the possible use of spatially continuous satellite cloud data for the purpose of analyzing the distribution of the dew point deficit by the method of optimum interpolation. There are two ways of applying this information.

The first is to use cloud data as supplementary information in regions with few aerological observation stations. For this purpose we must classify the cloud cover and find norms against which to judge mean values determined on the basis of selected dew point deficits for each cloud type. In objective analysis the values thus obtained are entered at the points of a uniform grid and in accordance with the satellite cloud pictures.

The second method consists in exposing certain statistical properties of the moisture field in its interrelation with the cloud distribution and to use these properties in the interpolation of values of the dew point deficit at points of a uniform grid. The field of cloudiness is here regarded as an indicator of the heterogeneity in the distribution of the dew point deficit. This will be explained in more detail.

As known, the method of optimum interpolation requires a knowledge of normalized autocorrelation functions and is applied only to homogeneous and isotropic fields /2, 4/. For this reason the

formation and dissipation of heterogeneity in the distribution of meteorological elements is of great importance. The first step in this
direction is an examination not of the fields of meteorological elements themselves, but of their deviations from normal. The second
step is the calculation of factors influencing the distribution of
normals of such characteristics as the annual variation in the examined elements, the geographical location of the observation points
and the character of atmospheric processes. Accounting for the
latter is rather difficult, but of great interest.

Cloudiness is a striking indicator of atmospheric processes.
The presence of large spaces occupied by uniform cloud cover indicates uniformity in the processes occurring there. Considering
the good correlation between cloudiness and relative humidity
/9, 10/, we can assume that the division of the field of dew point
deficit into regions with uniform cloud cover will bring us nearer
to conditions required for the calculation of normalized autocorrelation functions.

Objective analysis of the dew point deficit field by the method of
optimum interpolation with allowance for satellite cloudiness data
was conducted as follows.

Dew point deficit norms were obtained by analyzing some selected region with a homogeneous cloud cover with respect to stations
and points of a regular network. Interpolation of deviations from
these norms at the grid points is accomplished either at the limits
of each cloud-cover contour with the aid of corresponding normalized correlation functions, or, in the case of small differences
between these latter functions, in terms of the total field by means
of one averaged normalized autocorrelation function.

In the following we present the results of investigating the statistical structure of the field of dew point deficit at 850 mb, obtained
by analyzing data for April — March, 1967 — 1968.

For the computation of the normalized autocorrelation functions
according to cloud types we classified the cloud cover as viewed by
satellites. A rather detailed classification, including the amount
and thickness of clouds, was made by Thompson and West /10/.
Its application would be desirable, but was not possible in the first
stage of our work because the actually available satellite observations supply only TV photographs of the clouds in the visible part
of the spectrum without indicating their thickness.

Besides, only a detailed classification can significantly limit
the amount of data used in constructing the normalized autocorrelation functions of each type and requires much selection which,
up to now, it was not possible to undertake owing to insufficient
data. Therefore, in the basic classification we took only the cloud

amount and divided it into 5 main types: I — clear sky or thin upper-level clouds; II — small amounts of cumulus or stratocumulus clouds; III — considerable amounts of cumulus or stratocumulus; IV — continuous, nonfrontal cloudiness; V — continuous clouds of frontal origin.

When finding the normalized autocorrelation functions of the dew point deficit in the examined cloud masses we chose the region shown in Figure 1, as having the best coverage of satellite information and as being of greatest interest in weather forecasting for European USSR. The smaller rectangle limits the area in which, subsequently, an objective analysis can be made. Data from stations in the larger rectangle are used in interpolation over the territory of the smaller square. In computing the correlation functions data from all stations within the large square are used (about 200).

The choice of uniform cloud masses was made according to photos received from satellite ESSA, having a visual field of about 3000 km. The selected region was photographed by the satellite on four consecutive passes. The time of passage of the satellite over the eastern regions was 0900 Moscow time and over the western part about 1300 — 1400 hrs. Since the indicated region is better provided with radiosonde data for 1500 hrs Moscow time, we chose this observation time for the analysis. As a result the time difference between satellite and radiosonde observations in the eastern part of the examined region was 5 — 6 hours, but in the rest it did not exceed 4 hours. As the cloudiness is very variable the interval of 5 — 6 hours proved to be rather too large for making an analysis of the cloud cover in relation to the moisture field. However, regions with clear sky or with continuous frontal cloudiness which were much less prone to variability /7/ can be easily observed on the basis of synoptic maps during 5 — 6 hours. Consequently, a sufficient amount of data for statistically well founded conclusions was gathered only for types I and V. For this reason we give below data on the statistics of the dew point deficit at the 850-mb level only for clear weather and for frontal cloudiness.

Considering some time difference of aerological and satellite observations, the cloud data taken from mosaics were compared with surface cloud observations at points where there were radiosoundings. When the "clear sky" type on the mosaics corresponded to cloudiness of more than 3/8 of the sky according to surface observations, or when the "continuous cloudiness" type corresponded to clear sky or scattered clouds, these cases were excluded.

In selecting data for the same cloud type, data were chosen at an interval of at least two days.

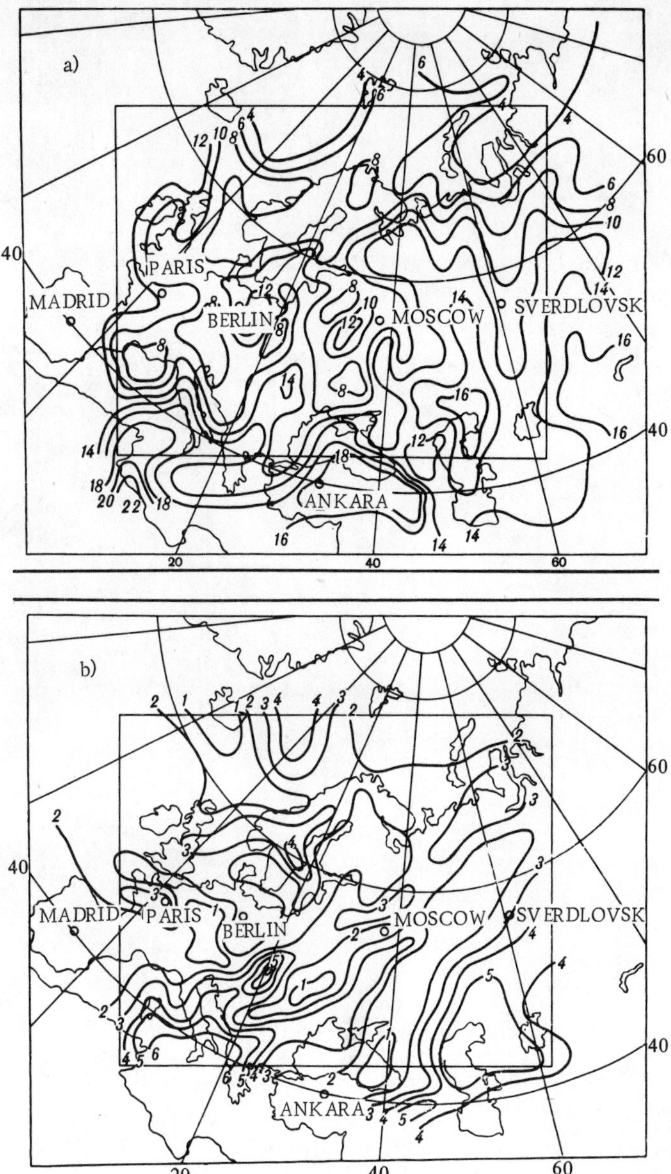

FIGURE 1. Distribution of dew point deficit norms at 850 mb:
a — for clear sky; b — for continuous frontal cloudiness.

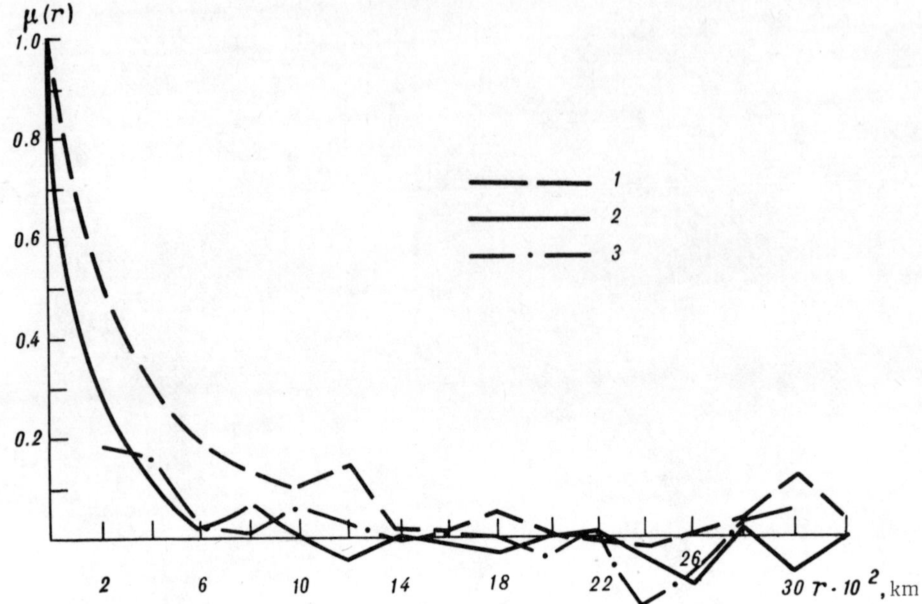

FIGURE 2.   Graph of the correlation of the dew point deficit at 850 mb:

1 — autocorrelation functions of the dew point deficit for clear weather; 2 — autocorrelation functions for continuous frontal cloudiness; 3 — cross-correlation functions of the dew point deficit with clear sky and with continuous frontal clouds.

The autocorrelation function of the dew point deficit for clear sky also decreases rapidly with increasing distance, but the correlation coefficient for clear sky is for all distances larger than that for continuous frontal cloudiness.   The radius of correlation for clear sky is 1400 km, i. e. approximately the same as for the autocorrelation function of the field of surface pressure tendency /3/.

Comparison of the autocorrelation functions of the dew point deficit with those of the temperature of the dew point, calculated without consideration of cloud data /5, 6/, shows that the point deficit possesses somewhat greater spatial variability.

As regards the cross-correlation function, there is no sense in dealing with it up to a distance of 400 km, since it was computed on the basis of insufficient data.   However, commencing from 400 km this function is hardly distinguishable from the autocorrelation function for frontal cloudiness, and from 600 km it is very close to zero, which gives evidence of the disappearance of the relationship at this distance.

Preliminary data on the above statistical structure of the dew point deficit field allow the following conclusions:

1) Since the dew point deficit for one type of cloudiness differs
considerably from that of another type in the objective analysis
carried out for the dew point deficit field, it is expedient to use
norms corresponding to the cloud type observed over the region of
interest.   The resulting differentiated norms will evidently be more
effective in regions with a sparse station network.

2) The obtained autocorrelation functions show that for interpolation of dew point deficit values at the points of a uniform grid
at the 850-mb level, one can use stations 200 — 400 km distant from
the grid points for frontal cloudiness and up to 600 — 800 km for
clear sky.

BIBLIOGRAPHY

1. Alpatova,R.L. Issledovanie statisticheskikh svyazei oblachnosti s defitsitom tochki rosy,
     vertikal'nymi skorostyami i temperaturoi (Investigation into the Statistical Relationship
     between Cloudiness and the Dew Point, Vertical Velocities and Temperature). — Trudy
     MMTs, No. 11. 1966.
2. Belousov,S.L., L.S.Gandin, and S. A.Mashkovich. Obrabotka operativnoi meteoro-
     logicheskoi informatsii s pomoshch'yu elektronnykh vychislitel'nykh mashin (Computer
     Processing of Meteorological Data). Leningrad, Gidrometeoizdat. 1968.
3. Belousov,S.L. and L.N. Strizhevskii. Sposob ucheta melkomasshtabnykh sostavlya-
     yushchikh pri ob"ektivnom analize polei baricheskoi tendentsii i prizemnogo davleniya
     (An Experiment in Calculating Small-Scale Components in the Objective Analysis of
     Pressure Tendency and Surface Pressure Fields).—Meteorologiya i Gidrologiya, No. 5. 1969.
4. Gandin,L.S. Ob"ektivnyi analiz meteorologicheskikh polei (Objective Analysis of Me-
     teorological Fields). Leningrad, Gidrometeoizdat. 1963.
5. Gandin,L.S.,V.V.Meleshko, and A.V.Meshcherskaya. O primenenii universal'nykh
     tsifrovykh mashin dlya issledovaniya statisticheskoi struktury meteorologicheskikh polei
     (Application of Universal Digital Computers to the Study of the Statistical Structure of
     Meteorological Fields). — Trudy GGO, No. 143. 1963.
6. Myach,L.T. O statisticheskikh kharakteristikakh i ob"ektivnom analize polya vlazhnosti i
     temperatury u poverkhnosti Zemli (Statistical Characteristics and Objective Analysis
     of Moisture and Temperature Fields at the Surface). — Trudy MMTs, No. 7. 1965.
7. Mazurin,N.I. Kharakteristika polei nekonvektivnykh oblakov i svyazannykh s nim zon
     osadkov (Characteristics of the Field of Nonconvective Clouds and their Associated
     Zones of Precipitation). — Trudy VNMS, Vol. 5. 1963.
8. Chuzavkova, N.A. O vozmozhnosti rascheta polya vlazhnosti po raspredeleniyu oblach-
     nosti (Possibility of Computing the Moisture Field from the Cloud Distribution). —
     Trudy Gidromettsentra SSSR, No. 11. 1967.
9. McClain,E.P. On the Relation of Satellite Viewed Cloud Conditions to Vertically In-
     tegrated Moisture Fields. — Monthly Weather Review, Vol. 94, No. 8. 1966.
10. Thompson,A.H. and P.W.West. Use of Satellite Cloud Pictures to Estimate Average
     Relative Humidity Below 500 mb with Application to the Gulf of Mexico. — Monthly
     Weather Review, Vol. 95, No.11. 1967.

FIGURE 1. Originals and quantified version of TV photographs:

1 — frontal cloudiness; 2 — clouds in a developing depression; 3 — "closed" convection cells; 4 — "open" convection cells; 5 — cumulus cloud rows; A — nonquantified originals against overlaid geographical grid; B — photographs, quantified with respect to five brightness grades: <0.20, 0.20—0.40, 0.40—0.60, 0.60—0.80 and >0.80 of the full range of the TV signal; C — photographs quantified with respect to three grades: <0.20, 0.20—0.40, >0.40; D — photographs quantified with respect to two grades: <0.20 and >0.20.

to know what particular objects are pictured in the photo, since a much simpler problem is dealt with: to define such parts of the picture in which similar objects are depicted. The second stage, recognition proper, consists of division of types of images used as complex decoding criteria to establish what actual objects (clouds, etc.) are depicted in each part of the photograph. Nevertheless, in different cases the same decoding criteria may possibly correspond to different cloud forms.

The programming, realizing an algorithm of decoding by two stages, comprises the following.

Let b_{ij} be the level of the TV signal corresponding to the i-th element of the j-th line of the TV photo. Then the matrix

$$
\begin{matrix}
b_{11}, & b_{21}, & \ldots, & b_{n1} \\
b_{12}, & b_{22}, & \ldots, & b_{n2} \\
\cdot & \cdot & \cdots & \cdot \\
b_{1n}, & b_{2n}, & \ldots, & b_{nn}
\end{matrix}
$$

characterizes in the computer memory the distribution of brightness (structure) of the image in part of the TV photo by $n \times n$ elements of resolution. The level of the TV signal for separate elements of resolution represents the first decoding criterion. However, its direct application for the determination of cloud forms is inconvenient because with every displacement, veer or change in the average level and range of the TV signal the values of b_{ij} vary independently from the change in the cloud images. The information value of these criteria is also very low owing to their strong correlation with adjacent elements of resolution.

An increase in the information value of these criteria may be obtained by several means. For decoding cloud images whose geometric outlines do not change, we assumed that it would be most expedient to use as criteria the coefficients of resolution of the photograph's brightness field with respect to some complete system of functions so that it can be used to represent every possible image. For simplicity in determining the likeness of different images the functions should be orthogonal, and for increasing the information value of the coefficients of resolution as indicators the accuracy of approximating the field of brightness must still be sufficiently high with a small number of terms of resolution. All these requirements are best satisfied by natural orthogonal functions

$$
\tilde{b}_{ij} = \sum_k x_k \varphi_{ijk},
$$

(latitude B, longitude L, azimuth A), and the orbit height and orientation of the optical axis of the TV cameras in relation to the satellite vertical as known.   According to the formula for central projection and transformation of coordinates one can find the latitude and longitude of the center of each picture part.   The transition of the coordinates of the map (ξ, η) is then accomplished according to formulas

$$\xi = \frac{1.8R}{m} \frac{\cos B}{1 + \sin B} \sin L,$$

$$\eta = \frac{1.8R}{m} \frac{\cos B}{1 + \sin B} \cos L,$$

where R is the earth's radius and m the map scale.   Then, according to the formula for meridians and parallels for the selected map projection we compute the coordinate of the geographical grid.   The geographical grid is set up on the alphanumeric printer of the electronic computer, and in locations corresponding to the centers of all recognized picture parts we print a previously agreed number — the recognition number of the viewed cloud type for each part.

Results of a comparison of automatic and visual nephanalyses of satellite TV pictures of the earth are given in Table 1.

TABLE 1. Results of comparison of automatic and visual nephanalyses of satellite TV pictures of the earth (number of cases, percent in parenthesis)

Visual nephanalysis	Cloudiness	Automatic nephanalysis			Total
		small	variable	considerable	
	Small	712 (50)	73 (29)	120 (11)	905
	Variable	559 (40)	148 (58)	254 (24)	961
	Considerable	135 (10)	34 (13)	697 (65)	866

Figure 2 shows one example of TV photographs recognized by the computer (Figure 2a), by visual nephanalysis with the input of surface observations of cloud amounts during the nearest synoptic time (Figure 2b), a map of types of TV images of this photograph constructed by the computer (Figure 2c), and a cloud map in three gradations: small — 00, variable — 01, and considerable — 02, also computer-prepared on the basis of the previously recognized TV image type (Figure 2d).

The given example shows sufficiently well the possibilities of our program for automatic processing and analysis of TV pictures. On this photograph the characteristic features of the cloud cover are seen: almost continuous cloud masses on the right central parts of the photo (over central and southeast Europe) and a large clearing over coastal western Europe and the regions surrounding the Baltic Sea. In the latter the photo shows the snow-covered parts of the continent. Over the North Sea convective clouds are developed. Variable cloudiness is seen in the Mediterranean. A narrow clearing shows over the Alps and on the upper Danube. The whole photograph was divided into parts of about 50×50 km^2. Each of these was analyzed in the computer memory separately, with the image type being recognized for each part and serving as input for the automatic nephanalysis.

In all, eight types were recognized and were determined by the size of the computer memory and by the desired rapid-action picture processing. In comparing Figures 2a and 2c one can see that the form of the recognized image types agrees well with the configurations recognized visually on the TV photograph. Even the cloudless parts of the land surrounding the Baltic Sea, covered by snow and therefore not distinguished in mean brightness from the adjacent cloudiness, was seen separated by distinct contours on the map of image types. The image of the snow-covered Alpine ridges was also distinguished by contours. Less successfully distinguishable are the convective cloud clusters over the North Sea, and this is easily explained. Clouds of this type are not characteristic for winter when the examined situation occurred. Therefore, the covariant matrix of the TV signal of winter photographs is not very suitable for describing the statistical structure of the image of cumuliform clouds. The number of these functions which describe the variation in the TV signal level of these clouds was too small, and this caused us to discard these functions as criteria in the image type recognition.

Figure 2d shows the result of estimating the cloud amount according to the three above-given gradations for each part of the photo on the basis of data of the image types recognized in these parts. One feature of our experiment was that, owing to the small amount of processed photographs, we did not succeed in obtaining estimates of a priori probabilities of observation for all three gradations of clouds on the basis of satellite observations. Instead we used corresponding surface-observed climatological values. Insufficient satellite and surface cloud observations were apparently the reason why it was not possible to distinguish the cloud-free snow-covered parts surrounding the Baltic Sea (Figure 2d). It

turned out to be included in "continuous clouds," although on the maps of image types the area was observed to be altogether clear.

On the whole, the result of automatic recognition is quite satisfactory, as can be seen by comparing Figures 2b and 2d, i. e., the visual and computer analyses. Obviously, these two analyses differ in detail, but a better coincidence of the configuration of separate cloud fields in the examined case and in other cases analyzed by us allow the conclusion that the processing program is quite convenient for large-scale analysis on the basis of satellite photographs. This is borne out also by the quantitative comparison of computer and visual nephanalyses in Table 1.

UDC 551.509.313

COMPUTATION OF HEAT FLUX IN NUMERICAL WEATHER FORECASTING

P. N. Belov, L. V. Berkovich, R. L. Alpatova

A method is described for calculating the radiative heat flux in numerical forecasting. Maps of radiative heat flux through the total atmospheric thickness are presented. The contribution of radiative heat flux to the diurnal temperature and pressure change is estimated. An example of a nonadiabatic forecast for surface pressure is given.

Results of numerical experiments in the computation of heat flux in numerical weather forecasting /1, 2/ have shown that it has a considerable influence on the forecasts. Its contribution to the diurnal actual pressure and temperature change is on the average 10—20%. However, in natural conditions this does not usually occur in a clear-cut form. In the atmosphere, apart from radiative, there are turbulent and phase change fluxes. Besides, it is remarkable that radiative and turbulent heat fluxes often almost entirely compensate each other. As a result the atmosphere is influenced only by their noncompensated parts, which depend, along with the latent heat flux, on nonadiabatic processes. This is explained by the fact that in the equations of heat flux the terms in the adiabatic parts are of a smaller order than the terms in the nonadiabatic parts. This allows one to neglect the adiabatic part in the solution of the problem of the temperature distribution in the atmosphere.

An experiment in designing a method of combined computation of radiative and turbulent heat flux was described by Berkovich /4/. At the beginning the problem to be solved was how to determine the standard temperature in the atmosphere as dependent on the mutual compensation of the main parts of radiative and turbulent heat flux. However, this method encounters serious difficulties in finding the thermal conductivity coefficients in the vertical, as determined by the distribution of the standard temperature.

In the present paper we propose another approach to the combined computation of radiative and turbulent heat flux. The standard part of the radiative heat flux which must compensate the standard part of the turbulent heat flux is calculated by the method

TABLE 2. Radiative heat flux throughout the whole atmosphere (deg/day)

19 January 1963

v	o	o	o	v	c	c	c	c	v	c
-1.8	-2.1	-2.1	-2.1	-1.4	-0.5	-0.4	-0.4	-0.4	-1.0	-0.6
v	v	v	o	v	v	v	v	v	v	o
-1.0	-1.9	-2.0	-2.1	-1.7	-1.4	-0.7	-0.9	-0.8	-1.2	-1.3
v	v	v	o	v	v	o	o	v	v	o
-1.3	-1.6	-1.6	-2.0	-1.5	-1.4	-2.1	-1.8	-1.4	-1.3	-1.4
v	v	v	v	v	c	v	o	v	v	v
-1.3	-1.7	-1.6	-1.6	-1.5	-0.9	-1.1	-1.4	-1.0	-1.5	-0.8
o	v	v	v	v	v	o	v	v	v	c
-1.3	-1.6	-1.4	-1.0	-1.5	-1.5	-1.2	-1.3	-1.2	-1.6	-0.7
o	v	v	v	v	v	v	c	v	v	v
-1.2	-1.6	-1.3	-1.2	-0.9	-1.2	-1.2	-0.5	-1.0	-1.6	-1.1
v	v	v	o	v	c	v	v	o	o	o
-1.2	-1.6	-1.5	-1.3	-1.0	-0.6	-0.8	-1.0	-1.3	-1.6	-1.4
c	v	v	o	o	v	v	v	v	o	o
-1.5	-1.6	-1.2	-1.5	-1.3	-1.3	-0.9	-1.1	-1.3	-1.6	-1.6
c	c	v	v	o	o	o	v	o	v	v
-1.5	-1.6	-1.3	-1.3	-1.2	-1.2	-1.1	-1.0	-1.1	-1.4	-1.4

28 June 1967

v	o	v	v	v	v	o	o	o	v	v
-1.5	-0.8	-0.7	0.0	0.0	0.2	-0.3	-0.4	-0.6	-0.7	-0.9
v	o	o	o	v	v	o	v	v	v	v
-1.0	-0.9	-0.7	-0.6	-0.6	-0.7	-0.7	-0.6	-0.7	-0.4	-0.6
o	o	o	o	o	v	v	o	v	v	v
-0.9	-0.9	-0.8	-0.5	-0.6	-0.4	-0.6	-0.5	-0.7	0.0	-0.4
o	v	o	v	v	v	v	v	v	v	v
-0.7	-0.4	-0.7	-0.5	-0.6	-0.8	-0.6	-0.6	-0.3	-0.2	-0.9
c	v	v	v	v	v	v	v	v	c	v
-1.3	-1.2	-0.5	-0.6	-0.5	-0.6	-0.8	-1.0	-0.8	-1.3	-0.5
v	c	v	v	v	v	c	c	c	v	v
-1.0	-0.9	-0.5	-0.6	-0.9	-0.9	-1.1	-1.1	-1.2	-1.1	-1.0
v	c	c	v	c	c	c	c	c	v	v
-0.9	-1.0	-0.9	-1.1	-1.0	-1.3	-1.1	-1.2	-1.3	-0.7	-0.7
c	c	c	c	c	c	v	c	v	v	c
-1.5	-1.3	-1.4	-1.4	-1.4	-1.4	-1.2	-1.3	-1.3	-1.1	-1.3
c	c	c	c	c	c	c	c	c	c	c
-1.5	-1.7	-1.5	-1.5	-1.4	-1.0	-0.5	-1.0	-0.8	-0.6	-1.4

TABLE 3. Standard radiative heat flux throughout different atmospheric layers (deg/day)

Layer, mb	January	July
0 — 100	-0.08	-0.02
200 — 500	-0.11	-0.05
500 — 700	-0.05	-0.54
700 — 900	-0.40	-0.41
900 — 1000	-1.17	-0.32

The numerical experiments of calculating the radiative heat flux were carried out with two forecasting models, a four-level quasigeostrophic model /1/ and a three-level model of complete equations /3/. The 1000, 700, 500 and 200 mb levels were involved in the first model and the 1000, 700 and 300 mb levels in the second.

The radiative heat flux was examined in two ways: in its total form (ε) and as a deviation from the standard flux (ε'). Estimates of forecasts were made by calculating the correlation coefficient (r) and the relative error (E) for the diurnal height change of isobaric levels.

To assess the influence of the radiative factor on forecasts, we also computed on the basis of the adiabatic model the values of Δr and ΔE, which are the respective deviations from the values of r and E for adiabatic and nonadiabatic forecasts. Positive Δr and ΔE correspond to better nonadiabatic forecasts in comparison with adiabatic, while negative Δr and ΔE correspond to worse forecasts. Table 4 gives some results of these experiments carried out for six situations.

Two of the examined six forecasts belong to winter situations and four to summer. In most cases the difference between adiabatic and nonadiabatic forecasts is small, especially if we consider the radiative heat flux in the form of the deviation ε'; this appears natural for one-day forecasts. The data in Table 4 show that it was successful in improving forecasts for 16 and 20 January 1963 for 300 and 700 mb in the model of the complete equations and for 16 January for 1000 mb in the quasigeostrophic model by accounting for the radiative flux in the complete form ε. Consideration of the flux in the form ε' lowers the success of the forecasts.

For summer forecasts the reverse picture appears: accounting for heat flux in the form ε' gives better results than in the form ε. In summer, at almost all levels a considerable deterioration is noted with the use of ε, whereas using ε' either improves them or there is only insignificant deterioration. From this we can conclude that in summer the noncompensated part of the radiative

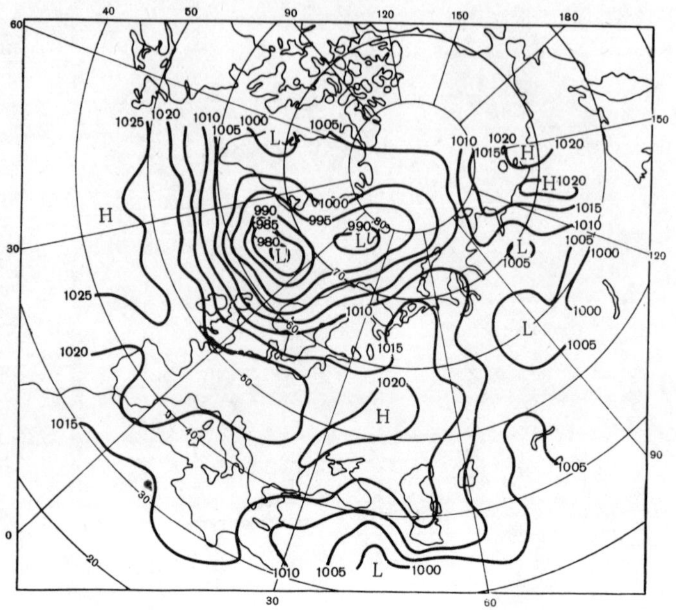

FIGURE 3.   Map of surface pressure at 0300 hrs on 29 June 1967.

heat flux plays an inferior role to that of the phase change and
turbulent flux.   The influence of these latter factors is considerably
less in winter, especially in the free atmosphere, whereas the radiative heat flux (cooling of the atmosphere) is most active.

In the forecasts for 29 May 1967 two kinds of data on clouds were
used when comparing the radiative heat flux: surface and satellite
data.   These experiments showed that the difference in the values
of radiative heat flux depend on the different cloud forms and were
found to be 0.1 and 0.2 deg/day, and the change in the correlation
coefficient and the corresponding error was 0.02 − 0.03.   Therefore,
in the comparison of radiative heat flux, surface cloud observations
can be placed by satellite data in such experiments.

As an example, a nonadiabatic prognosis is given for 29 June
1967.   Figures 1 − 3 show the initial, forecast and actual maps of
surface pressure.   The main process in this example is the intense
depression over the Norwegian Sea, where the surface pressure drop
reached 23 mb/day. At the same time the pressure change in the
center of the depression near Iceland was 6 mb.   The depression
was displaced 800 km to the northeast in one day.   On the prognostic
map the center of the depression almost coincides with the location
of its center on the actual map, but the pressure drop is only 15 mb,
i. e., 8 mb less than the actual change.

The intensification of the ridge over north European USSR and filling of the depression over the western Baltic Sea are quite well marked on the prognostic map. The relative error in forecasting the surface pressure in this example and the correlation coefficient are 0.60 and 0.79 respectively, according to the model of complete equations. The difference between the estimates of the nonadiabatic and adiabatic forecasts of the surface pressure is 1% (Table 4) and the pressure change caused by the radiative heat flux is about 1 mb at various points.

BIBLIOGRAPHY

1. Belov,P.N. and R.L. Alpatova. Nekotorye rezul'taty chislennykh eksperimentov po uchetu radiatsionnogo pritoka tepla v chislennom prognoze (Some Experiments in Using the Radiative Heat Flux in Numerical Forecasting). — Trudy Gidromettsentra SSSR, No. 11. 1967.
2. Belov,P.N. and A.F.Kivganov. Ob izmeneniyakh temperatury i geopotentsiala, obuslovlennykh radiatsionnym pritokom tepla (Changes in Temperature and Geopotential Caused by the Radiative Heat Flux). — Trudy MMTs, No. 8. 1965.
3. Berkovich,L.V. Trekhurovennaya skhema prognoza meteoelementov po polnym uravneniyam (A Three-Level Forecast Model for Meteorological Elements According to the Complete Equations). — Trudy MMTs, No. 14. 1966.
4. Berkovich,L.V. O sovmestnom uchete radiatsionnogo i turbulentnogo pritokov tepla v chislennom prognoze (Combined Calculation of the Radiative and Turbulent Heat Flux in Numerical Forecasting). — Trudy Gidromettsentra SSSR, No. 11. 1967.
5. Voitko,N.V. and A.F.Kivganov. Raschet i analiz radiatsionnykh potokov i pritokov tepla pri nalichii oblachnosti (Calculation and Analysis of Radiative Heat Fluxes and Heat Flow in the Presence of Clouds). — Trudy MMTs, No. 11. 1966.

For this reason one can assume that by measuring the radiation in the $8-12\mu$ spectrum one can determine sufficiently accurately the real temperature of the tops of dense, continuous clouds (averaged over the field of scan of the instrument). If the vertical temperature distribution in the atmosphere is known one can use information on the temperature of the cloud tops to find their height.

In this paper the actual vertical temperature distribution in the atmosphere was assumed unknown. The problem was to use only mean climatic data on the state of the atmosphere in the interpretation of satellite observations to reveal the satellite as a suitable observation platform for regions with sparse meteorological surface observations.

In using mean climatic data a number of (additional) errors arise in the determination of the cloud top heights. These errors are caused by the variation in the vertical temperature profile relative to the average distribution, and by the variable water vapor content of the atmosphere. The latter is not too large, and in the case of clouds with rather high tops $(4-5$ km) it may be neglected. Variations in the vertical temperature profile are more important. As shown by computations $/2-4/$, the values of these variations vary for different atmospheric layers and depend on the geographic position and season. The maximum standard deviation of temperature is observed near the surface and in the tropopause layer. For the same region of the globe the temperature variation is larger in winter than in summer. The average variation in temperature is somewhat greater than that of summer. One can say that the maximum variation (standard deviation) of temperature is of the order of $10°$. Its most frequent value is $3-6°$. One may therefore expect that the mean error in the determination of the cloud tops computed on the basis of climatological means of the vertical temperature distribution will vary within $0.5-1.0$ km.

A scheme for determining cloud top heights by computer has been planned, using the radiation temperature in the $8-12\mu$ interval of the spectrum, referring to space and time and expressed in degrees Celsius.

The globe was divided into $5°$ latitude and $10°$ longitude rectangles. For each of these the climatological mean temperature and geopotential were used as input into the computer at some standard isobaric levels, as well as the corresponding factor ΔT for the radiation temperature according to climatological data on the distribution of temperature, moisture and ozone for a given region. Four standard isobaric levels were chosen. The actual choice of these levels may be different. Usually, for many computations of temperatures and geopotential, data for the 850, 700, 500 and 300 mb levels were chosen.

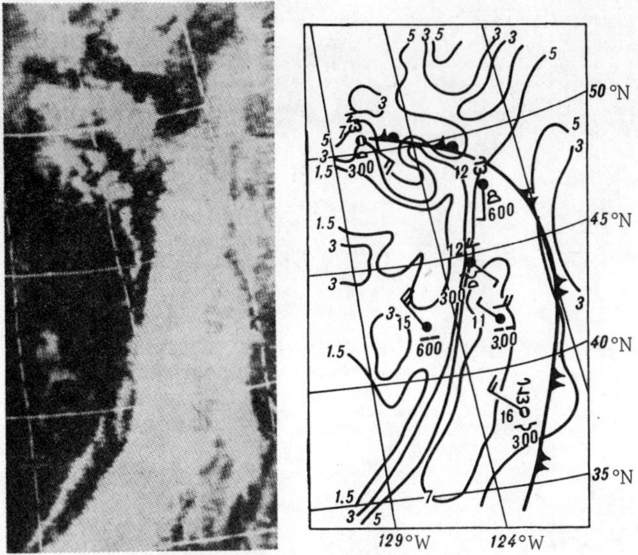

FIGURE 1. Example of a map of cloud top heights (in km) for
an actual synoptic situation.

Computation of cloud top heights and compilation of the maps
were made by using data from Cosmos 226 and Meteor satellites
(1968 — 1969).

The cloud top heights were presented in map form, similar to the
matrix charts used to display radiation data.

Comparison of maps of cloud top heights with infrared photographs obtained by high resolution radiometers showed that there
is good agreement between these two forms of satellite information (Figure 1). The forms of (smoothed) cloud top height isolines
repeat in general the contours of the cloud distribution as seen on
the infrared pictures.

The compatability of data obtained by the two independent func-
tionings of the satellite radiometers allows one to limit the processing of digital data required in decoding infrared pictures received from measurements of medium resolution radiometers.

It is also of interest to have some quantitative estimates of the
accuracy with which cloud top heights are determined by using data
from independent observations, such as surface synoptic meteorological stations. The most extensive data on the state of the free
atmosphere are nowadays obtained from radiosonde observations.
These, however, give no accurate values of cloud top heights and
contain only indirect information on the vertical distribution of

the compared values for layers of 1 km thickness (except for the
0 — 2 km layer),there is reason to believe that for certain cloud
types the height of their tops can be determined with the same accuracy using synoptic methods.

Consequently, an operational scheme for determining the height
of cloud tops, using as supplementary information only climatological
mean distributions of temperature and moisture, allows one to obtain
quantitative data on the relief of the upper limits of quite thick,
horizontally and vertically extensive clouds.

BIBLIOGRAPHY

1. Boldyrev,V.G. Raschet peredatochnykh funktsii atmosfery v intervale, 8 — 12 mk dlya
      territorii severnogo polushariya (Computation of Transmission Functions for the Atmo-
      sphere in the 8—12μ Interval for the Northern Hemisphere). — Izvestiya AN SSSR,
      Seriya Fizika Atmosfery i Okeana, Vol. 1, No. 7. 1965.
2. Boldyrev,V.G. O mashinnom postroenii kart temperatury zemnoi poverkhnosti i vysoty
      verkhnei granitsy oblakov po dannym meteorologicheskikh sputnikov Zemli (Computer-
      Constructed Temperature Maps of the Earth's Surface and Cloud Top Heights from
      Meteorological Satellite Data). — Trudy MMTs, No. 11. 1966.
3. Koprova,L.I. and V.G.Boldyrev. O statisticheskikh kharakteristikakh vertikal'noi
      struktury polei temperatury i vlazhnosti do bol'shikh vysot (Statistical Properties of the
      Vertical Structure of Temperature and Moisture Fields up to Great Heights). — Izvestiya
      AN SSSR, Seriya Fizika Atmosfery i Okeana, Vol. 6, No. 2. 1970.
4. Stel'makh, F.N. Vertikal'nye korrelyatsionnye svyazi temperatury v svobodnoi atmosfere
      nad severnym polushariem (Vertical Temperature Correlations in the Free Atmosphere
      over the Northern Hemisphere). — Trudy NIIAK, No. 30. 1965.

UDC 551

FEATURES OF THE EVOLUTION OF DEPRESSIONS AND THEIR CLOUD SYSTEMS OVER THE PACIFIC OCEAN

Yu. V. Kurilova, I. R. Egorova

Satellite radiation maps and nephanalysis charts are used to investigate the evolution of the cloud field during the development of depressions.

The development of depressions and of the cloud field is examined as a function of large-scale features of the pressure field. Especially emphasized is the role of blocking ridges. Features of the distribution of cloud fields in the presence of latitudinal cyclonic transformations are revealed.

This paper deals with the evolution of cloud systems in relation to the development of depressions with the aid of satellite radiation measurements and nephanalysis maps. The region chosen for this investigation was the eastern Pacific Ocean from 160°E to 100°W and from 10°N to 70°N, which has the best coverage of satellite radiation measurements.

The evolution of cloud systems belonging to depressions was studied in two continuous series: 12 − 20 January and 3 − 21 February 1968. This choice of material allowed the passage of one depression to be tracked over the investigated area during several days. The succession of cyclonic perturbations was studied as they passed this region throughout the above-mentioned period. In January the evolution of four depressions and their cloud systems was followed, and in February that of six such depressions. The development and displacement of the depressions were investigated in connection with large-scale pressure fields over the Pacific.

In the analysis schematic maps were used for every day of the series. On these maps the cloud systems were plotted on the basis of nephanalysis, −40 and +10° isotherms of the radiation temperature, the position of fronts, surface and upper-level pressure centers, the position of axes of ridges and troughs and 580 decameter isohypses.*

---

* Regions of low radiation temperature, as delimited by the -40° isotherm, shall in the following be called minima of radiation temperature, or radiation minima.

Examples of such maps for the first four days are given in Figure 1. It was not possible to present maps for the remaining days, and so Figure 2 shows composite maps for January and February 1968 of the axes of ridges, trajectories of the consecutive surface depression centers $(D_a, D_b, D_c, D_d, D_e)$, the positions and orientation of radiation minima (whenever available).

Some basic parameters taken from the schematic maps are given in Table 1.

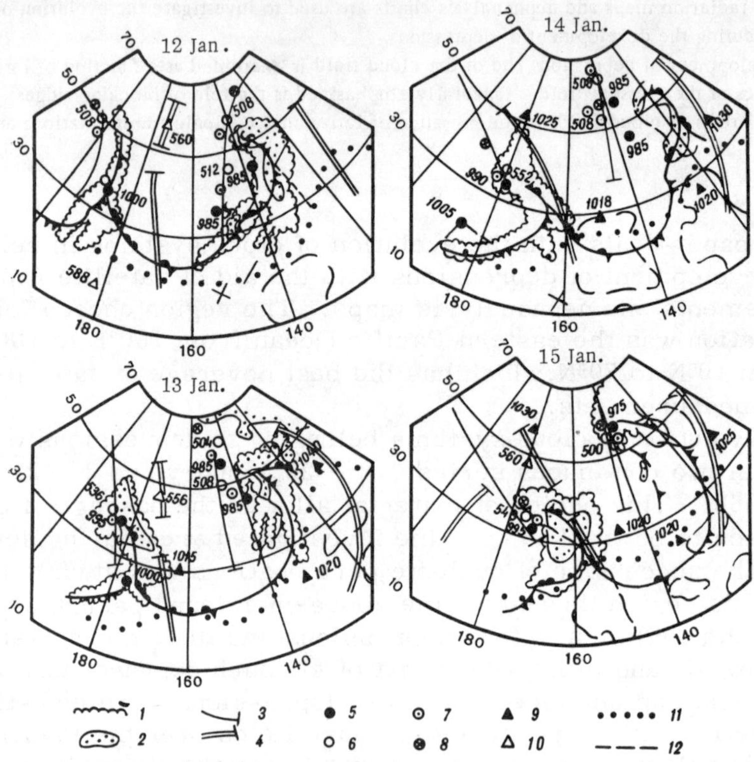

FIGURE 1. Schematic maps of frontal cloud systems and pressure fields for 12—15 January 1968:

1 — continuous clouds according to nephanalyses; 2 — region of radiation temperature below -40°; 3 — axes of troughs; 4 — axes of ridges; 5 — surface depressions; 6, 7, 8 — regions of low values of AT_{500}, AT_{700} and RT_{1000}^{500}; 9, 10 — surface and upper level anticyclones; 11 — 580 decameter isohypse; 12 — 10° isotherm of radiation temperature.

1. CHARACTER OF LARGE-SCALE PRESSURE FIELDS AND THE EVOLUTION OF CLOUDINESS ASSOCIATED WITH THE DEVELOPMENT OF DEPRESSIONS IN JANUARY 1968

The main features of synoptic processes in January 1968 were the simultaneous existence of a persistent upper level ridge at the American coast and of another ridge in the central Pacific which sporadically appeared and disappeared (Figure 2).

In the period 12—15 January a depression with two centers (D_a) was located between these two ridges. This low reached its greatest intensity on 15 January (Figure 2, Table 1). About the same time (14 January) a closed cold pocket formed in the rear of the depression. Considerably later, in the transition stage from a growing to a mature depression, the occlusion began; this explains the cloud band oriented almost meridionally. The triangular form, which is typical for the growing cyclone stage, maintains only a mass of high clouds up to the formation of a closed cold center on 12 and 13 January. After the appearance of this closed pocket the high clouds also assume a spiral form.

The orientation of the areas of low radiation temperature relative to the depression center is in this stage to the east of the surface pressure center: at first SSE and then NNE of it.

The height and thickness of the cloud masses, in agreement with the lowest radiation temperatures (Table 1), increased somewhat with deepening of the depression from 12 to 14 January (the radiation temperature decreased from —52° to —62°), and at the moment of maximum deepening of the depression the height of clouds began to decrease (the radiation temperature rose to —50°). The area occupied by high clouds and surrounded by the —40° isotherm was largest on 12 January ($12 \times 10^5 \, \text{km}^2$), after which it was gradually reduced, so that on 15—16 January the high clouds occupied only $3 \times 10^5 \, \text{km}^2$ to the north of the occlusion (see Figure 1).

After 16 January the cloud system of depression D_a already ceased to exist independently, but the cloud system of the following two depressions, which were of small intensity (D_b and D_c), merged with it after arriving from the west. The evolution of the thickness of these clouds of the first depression was estimated according to the radiation temperature as follows. In the stage of deepening of the depression, when it was blocked by the anticyclonic ridge, the vertical thickness of the cloud mass increased ($T_r = -45°$ on 13 January and $T_r = -65°$ on 15 January). With the low's further deepening during its eastward motion and its merging with the old filling depression the cloud thickness remained almost unchanged

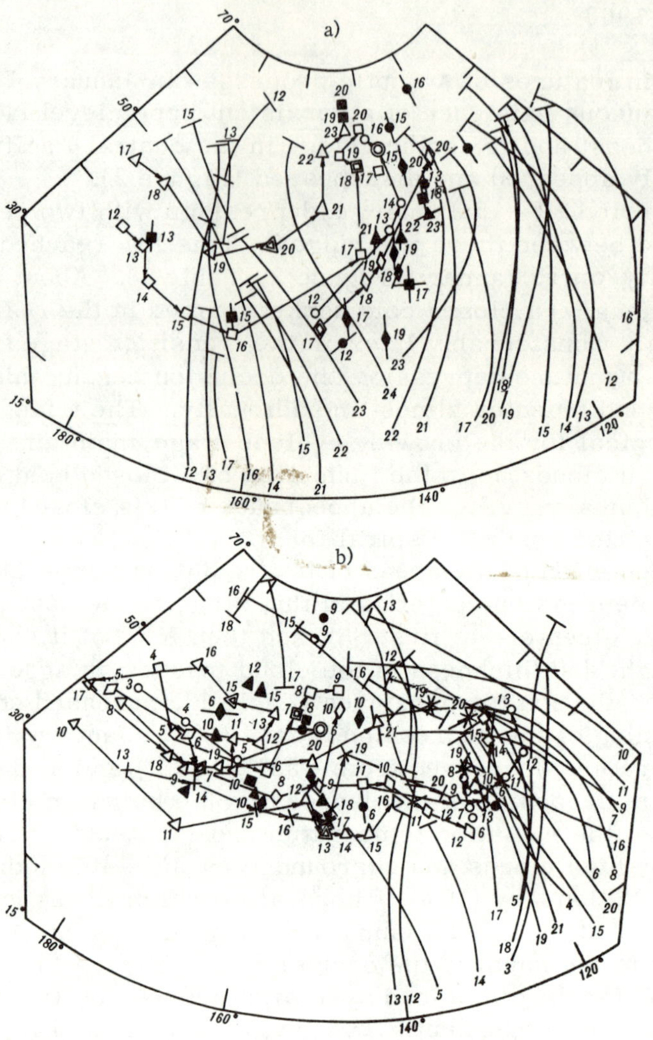

FIGURE 2. Composite maps of surface depressions corresponding to
Table 1 (open symbols 1 — 5, double symbols — deepest depressions),
axes of blocking ridges (thin lines), and minima of radiation temperature (black symbols):

a — January; b — February; 1 — D_a; 2 — D_b; 3 — D_c; 4 — D_d;
5 — D_e.

TABLE 1. Evolution of pressure centers and associated changes in radiation parameters

January 1968

Number	D_d						D_c						A	D_b					
	1	2	3	4	5	6	1	2	3	4	5	6	7	1	2	3	4	5	6
12																508	1000		
13														+		536	995 1000	—45	5
14														+	+	552	990		
15														+	+	548	995	—65	10
16													53	+	+	552	990		
17			520	985					540	990			30	+	+	540	995	—62	11
18	+	+	532	985					532	995				+	+	500	985	—63	15
19	+	+	520 524	985 985			+		536	995	—62	5					995	—53	30
20	+	+	516 520	970 970						1005	—61	3					1005	—58	20
21				975 980	—48	20													
22				975 1000	—60	17													
23				1000 1005	—60	14													
24				995	—53	2													

TABLE 1 (Continued)

January 1968

Number	A			D_a						A			D_{upper}
	7	8	9	1	2	3	4	5	6	7	8	9	3
12	63 44	—20	—30	+	+	508 512	985 985	—52	12	52	—10	—20	
13	57 34	—10	—20	+		504 508	985 985	—53	8	55	0	—20	
14	54			+	+	508	975	—62	7	58	5	—10	
15	60 40	—10	—25	+	+	500	975	—52	4	53	0	—15	
16				+	+	504	985	—50		51	0	—15	
17						500	980			47	—5	—20	
18										52	—10	—20	
19										54	0	—20	564
20										55	0	—20	560
21										55	0	—20	560
22										59	0	—20	572
23										55	—10	—20	576
24										46			576

TABLE 1 (Continued)

Number	D_d				A	D_e						D_d						D_c					
	1	2	3	4	7	1	2	3	4	5	6	1	2	3	4	5	6	1	2	3	4	5	6

February 1968

Number	D_d 1	2	3	4	A 7	D_e 1	2	3	4	5	6	D_d 1	2	3	4	5	6	D_c 1	2	3	4	5	6
3																							
4																							
5																							
6																				516	980	—52	2
7																				512	980		
8															1000 1000					532	985	—42	1
9												+		516	995 980	—42	2				995	—41	1
10												+		504	970	—61 —51	5 4	+			1000		
11												+	+	516	975 980			+			1005		
12												+	+	508	975 970	—46 —38	6 10						
13						+		504	970			+	+	500	955 985								
14		+	504			+	+	508	965			+	+	548	985								
15	+	+	504	965		+		548	995			+	+	544	985								
16		+	500	965		+		540	980	—43	1	+		540	990	—40	1						
17	+		508	970		+		520	975														
18	+	+	524	970	32	+	+	532	980														
19	+		524	985	35	+		524	980	—45	6												
20	+	+	516	980	43	+		520	980	—60	7												

TABLE 1 (Cotinued)

Number	D_b						D_a						A			D_upper	
	1	2	3	4	5	6	1	2	3	4	5	6	7	8	9	3	3
									February 1968								
3							+		508	985			46	0	—15		
4			512	995			+		516	980			50				
5			520	980			+	+	520	975			45	—6	—10		
6			508	965	—44	1	+	+		975	—52	5	45	—10	—20		564
7	+		508	970	—46	6							54	0	—10		560
8	+	+	512	980	—48	2							55				560
9	+	+	516	990									61	—20	—30		560
10				995	—42	1							60	—18	—30		556
11													68			540	556
12													67	5	—10	556	556
13													70			552	
14													68	—20	—30	540	
15													68			544	
16													43	—20	—30		
17													47				
18													44				
19													49				
20													49	—10	—30		

N o t e. In columns 1 and 2 the plus sign signifies the beginning of occlusion and a closed cold pool, respectively, on the map of the relative topography (RT); columns 3 and 4 give the height and pressure of the upper and surface pressure centers, respectively; columns 5 and 9 give the minimum radiation temperature; column 6 gives the area outlined by the -40° radiation temperature isotherm ($\times 10^5$ km); column 7 gives the latitude of the 556 decameter isohypse; column 8 gives the maximum radiation temperature. Individual series of depressions are distinguished by lines.

$(T_r = -62$ to $-63°)$, whereas in the filling stage of the surface depression and its northward displacement on 19 and 20 January it decreased $(T_r = -53$ to $-58°)$.

The area of high clouds marked out by the $-40°$ radiation temperature isotherm gradually increased during the development cycle of this depression from $5 \times 10^5 \, km^2$ on 13 January to $30 \times 10^5 \, km^2$ on 19 January, and it began to decrease only on the last day of the cycle (Table 1). The maintaining of such a large area is explained by the merging of the cloud system of depression D_b with that of the old depression D_a. A similar widening of the cloud mass during merging was also noted with depression D_c.

In the first days, from $12-15$ January, during the deepening stage, the cloud masses stretched out vigorously in a meridional direction along the axis of the ridge, corresponding to the occlusion; on 15 and 16 January, when the blocking ridge advanced eastward, the cloud system, as shown by the nephanalysis and also by the radiation map, took on the triangular form, characteristic for the growing stage of a depression. The appearance of occlusion before the formation of a closed cold center can be attributed to the effect of a stable blocking ridge.

When on 17 January the cloudiness of the examined depression D_b merged with that of the preceding depression D_a it formed a vast, irregular cloud band which already began to narrow down in the last days. A remarkable feature of the breaking up of these clouds was the appearance of small isolated centers outlined by the $-40°$ isotherm, observed on 20 January. The new depression D_c, in contrast to the above examined, which came from the west, formed in situ on 17 January, splitting the blocking ridge into a northerly and a southerly part. As mentioned later, this depression formed in the region where cloudiness was persistent (Figure 3) under the influence of the blocking ridge.

The following cyclonic center D_d moved during its deepening from Kamchatka eastward, and advanced north during filling from 21 January onward, while approaching the ridge (Figure 2a).

At the beginning of its filling stage (21 January) the cloud system of this depression was very large (about $20 \times 10^5 \, km^2$) owing to its merging with the old cloud system.

The stage of decay and dispersal of the cloudiness was well pronounced, because the advance of new depressions from the west stopped, owing to the development of a blocking ridge over the ocean from 21 January.

The decay of cloudiness began with narrowing of the cloud band on 22 January $(17 \times 10^5 \, km^2)$. On 23 January a thick cloud band split apart, and on 24 January it was reduced to two small patches

$(2 \times 10^5 \, km^2)$. The height of tops of the cloud system, which was oriented in a meridional direction to the west of the ridge during filling of the depression (on 22 and 23 January), was still considerable ($T_r = -60°$). The tops began to drop in height only with the breaking up of the cloud mass ($T_r = -53°$).

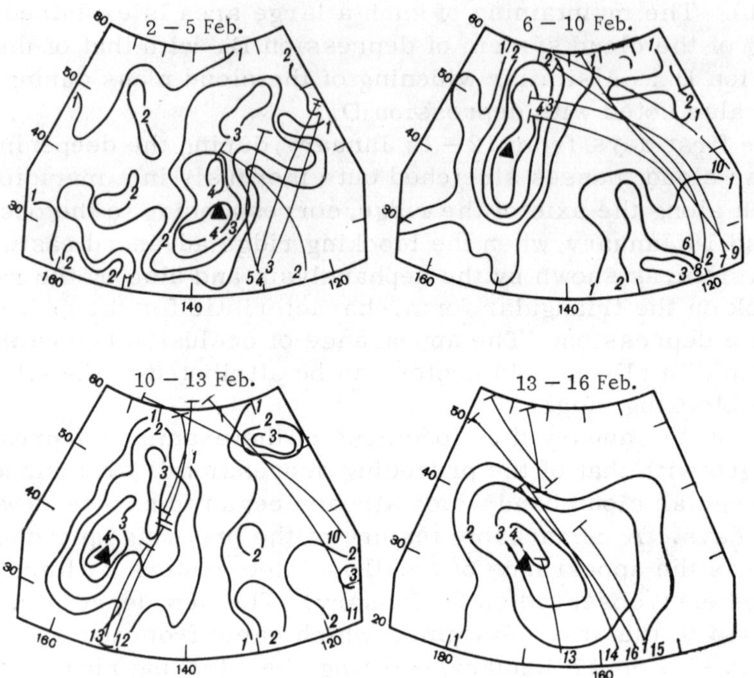

FIGURE 3. Map of the number of days with cloudiness for the four days preceding the appearance of an upper level low (black triangles) and the position of axes of the blocking upper ridges (thin lines).

It is seen from the above example that an upper level ridge delays the advance of cloud systems and causes them to become stationary to the west of the ridge.

When there are two ridges the clouds are usually concentrated between the axes of these ridges and the axes of the troughs (to the west of one ridge and to the east of the other). The western ridge inhibits the advance from the west of new cloud systems and creates conditions for a full cycle of development, including disintegration and dispersal of cloud systems blocked by the eastern ridge. In this way, under a meridional upper level circulation to the east of troughs and to the west of the ridge axis there is always cloudiness, while to the east of the ridge the sky is cloud-free.

2. CHARACTER OF THE LARGE-SCALE PRESSURE FIELD AND EVOLUTION OF CLOUDINESS IN RELATION TO THE DEVELOPMENT OF DEPRESSIONS IN FEBRUARY 1968

As can be seen from the trajectory maps (Figure 2) the synoptic processes over the east Pacific Ocean were zonal during February, and all cyclonic centers lay within 45 — 50°N. A deviation from zonality occurred only at the west coast of Central America owing to a persistent orographic ridge and the coast line (Figure 2, Table 2). In the first five days the ridge was only forming and lay south of 50°N. During its further development it moved north, and toward the end of the second five-day period it reached 60°N, while in the third five-day period it reached its greatest development and extended to 70°N. After a transformation to a more cyclonic flow appearing in the fourth five-day period the ridge retreated to 40°N.

This character of the upper level pressure field as established in February favored conditions for the repeated eastward motion of ever new depressions coming from the west. Thus, for a 20-day period one could follow the movement of depressions of various intensity over the analyzed region of the Pacific. Since these depressions were mostly formed in the west Pacific in a persistent frontal zone they usually underwent a larger or smaller development in the investigated region (east of 180°).

The development of one of the most active depressions (D_a) can be traced from 3 February, when two additional centers $(D_b$ and $D_c)$ formed in it.

The cloud system of the young depression had on 3 February the shape of an elongated triangle whose two sides corresponded to the cold and warm front, while the meridionally oriented and longest side was formed by the occlusion. On 4 February the cloud band was still more meridionally extended, assuming a spiral form. This example also shows that the occlusion starts in the stage of deepening of the depression (during the two days before reaching its maximum depth) and until the formation of a closed cold core in its rear.

In the stage of maximum deepening and the formation of the cold core on 5 and 6 February the height of cloud tops, according to the radiation temperature $(T_r = -52°)$, remained considerable, although its area decreased somewhat.

On 7 and 10 February the cloud systems of two consecutive nonactive secondary depression centers $(D_b$ and $D_c)$, arriving from the west and not developing independently, merged with the principal cloud system. The first of these centers (D_b) appeared over the examined region on 4 February (512 decameters — 995 mb), and on 6 February it formed a young depression (508 decameters — 965 mb) with a relatively small cloud mass of circular form ($T_r = -44°$).   Toward 8 February this depression occluded and its cloud system, approaching the belt of the previously occluded depression, formed a complex Π-form system, and on the following day these two fronts merged into one broad belt.

TABLE 2.   Velocities v of movement (deg. latitude/day) and distances r of depressions from the axes of upper level blocking ridges (deg. latitude)

January 1968

Number	D_d		D_c		A	D_b		A	D_a		A
	v	r	v	r	λ	v	r	λ	v	r	λ
12						12	25	170	5	32	115
13						7	11	170	2	31	115
14						5	13	180	2	40	110
15						12	17	180	—2	40	110
16	—2	25			175	6	10		—4	45	100
17	4	25	12	52	165				1	30	125
18	27	70	18	55							120
19	12	60									115
20											120

February 1968

Number	D_d		A	D_e		D_d		D_c		D_b		D_a		A
	v	r	λ	v	r	v	r	v	r	v	r	v	r	λ
3												25	80	115
4										12	36	22	20	122
5								12	42	12	30	0	6	120
6								12	40	6	22			140
7								12	36	4	22			140
8								4	24		24			138
9						18	60							140
10						6	30							140
11						2	30							140
12				20	60	2	40							140
13				18	50	30	40							140
14				14	40	4	12							140
15			150	6	34	0	16							140
16	—15	10	160	4	16									127
17	2	25	165	4	26									127
18	20	70	180	5	24									120
19	22	60												123
20	8	40												115

The second shallow cyclonic center (D_c) was first noted in the radiation temperature field on 6 February as a cold area ($T_r = -52°$), having an elliptic form and extending latitudinally.   As the depression filled, its cloud field changed on 9 February into a very small,

meridionally oriented band of low clouds ($T_r = -41°$) which on
10 February merged with the cloud system of the preceding depression.

The development of the active depression D_d, which had two
centers, was considerably more interesting. It underwent a complex
development on 8 February: a deepening, blocked and retrograde
motion of the northern center and a simultaneous separation and
southeast movement of secondary and shallower depressions. On
9 February it reached the stage of a young depression with a relatively low cloudiness in which only near the occlusion point the
radiation temperature was below $-40°$. As in the preceding cases,
the depression had begun to occlude already with the formation of
an isolated cold pool and at the stage of lowest pressure.

It is interesting to note in this example the fluctuation in width
and height of the cloud bands in relation to the fluctuation of the
lowest pressure of the depression. When the surface low reached
its first minimum on 10 February (504 decameters — 970 mb) and
the cyclonic circulation extended up to 100 mb, the cloudiness was
arranged in the form of a sharply twisted spiral band of high clouds
($T_r = -61°$).

The partial filling of the low on 11 — 12 February and the lowering
of the level of cyclonic circulation to 300 mb on these days was accompanied by narrowing of the cloud band and lowering of its upper
limits: the cloud mass was outlined by the $-30°$ isotherm (Table 1).

A second deepening of this depression on 13 February (500 decameters — 955 mb) was accompanied by a renewed broadening of
the cloud band, as seen on the nephanalysis.

The renewed filling of the low, following 14 February, was connected with the separation from the main depression of a secondary
upper-level cyclonic center, causing breaking up of the cloudiness.
This was especially manifest in clouds with a flaky appearance on
15 February.

However, the dissolving cloud band continued to persist to the
west of the ridge for several days, till 19 February, even after the
disappearance of the surface low and its fronts. This again confirms the great persistence of cloud systems.

The principal depression became a secondary low from
14 February under the influence of a blocking ridge extending from
northern latitudes, and then moved erratically, being pushed to the
west from 160°W to 160°E. From 17 February the low began to
shift again eastward in the usual way, closing the loop described
by its trajectory (Figure 2). Its cloud system had a form typical
for a growing system, i. e., the meridionally extending triangle
shown at the beginning of occlusion.

In the further development and eastward motion of the depression the northerly part of its cloud system was blocked by the above-mentioned ridge from the north, whereas the southerly part was clearly displaced eastward. This favored a peculiar cloud configuration: a narrow band of occlusion extended strongly from northwest to southeast and a cold frontal band from northeast to southwest.

After the dissolution of the anticyclone in the north (on 19 February) the entire cloud system clearly moved east up to the blocking anticyclone at the North American coast and reached it on 21 February.

At the same time as depression D_d moved back to the west, another depression D_e moved east. This low at first deepened up to 14 February and then began to fill partly (15 and 16 February). Its cloud system also had the typical triangular contours in the growing stage, with one very elongated side corresponding to the cold front.

When this depression reached a second minimum on 17 February, its cloud system assumed the form of a narrow spiral band, typical for an occlusion.

This cloud band approached together with the preceding cloud band, the blocking ridge, and on 20 February already merged with it, forming one vast cloud mass whose tops grew somewhat after the merging ($T_r = -45$ to $-60°$).

3. ROLE OF BLOCKING RIDGES IN THE MOVEMENT OF DEPRESSIONS AND THEIR CLOUD SYSTEMS

We shall dwell especially on the effect of blocking upper ridges on the motion of depressions and their frontal and cloud systems. In the example of the development of 10 analyzed depressions in January and February 1968, it was seen that the velocity of movement of upper-level lows changes markedly, depending on the position of blocking upper ridges. Table 2 shows that diurnal velocities of displacement of depressions located at more than 60° longitude from the axis of a ridge move with a velocity of about 20° latitude per day. When they approach to within 60 − 30° latitude of the ridge axis their mean velocity changes to 8°/day, and for distances of less than 30° it becomes only 2°/day.

This is much slower than the generally accepted velocity of tropospheric perturbations which, in /4/, was taken as 35 − 45°/day for a zonal flow. Consequently, at a distance of 60° the ridge

already begins to exert a braking influence on the advance of
depressions.

Let us choose several examples of the movement of upper lows.
Depression D_a slowed down between 3 and 6 February from 25 to
0° latitude/day (Table 2). Depressions D_b and D_c, when merging
with the preceding depression, conserved their velocity of about
12°/day for 2 — 3 days while approaching to within 36 — 40° of the
ridge. With further approach up to 24° these lows slowed down
to 4°/day. The following depression D_d slowed down to 2 — 6°/day
already when it was 30 — 40° away from the ridge. In this case the
braking effect was caused by the upper part of the ridge, which
curved latitudinally around a low that had separated from the main
low.

With the appearance of this particular cyclonic center there
occurred a sharp discontinuity in the velocity of the low from 2 to
30°/day on the eve of 13 February.

Such a clearcut and regular decrease in the velocity of depressions during their approach to a blocking ridge could be observed
also with depression D_e and the second center of depression D_d
when, after having regressed westward, it began to move again
to the east.

Thus, the dynamic head represented by the blocking ridge strongly
influences the velocity of depressions and, we may say, governs their
motion.

4.   EVOLUTION OF PRESSURE SYSTEMS AS REFLECTED
BY RADIATION PARAMETERS

Below we summarize existing, though uncoordinated radiation
measurements for the examples examined in paragraphs 1 and 2.
The purpose of this summary is to try and find some parameters
for assessing the character of development of pressure systems
with respect to the radiation field.

As pointed out, between the axis of the trough (to the east of it)
and of the ridge (to its west) there usually exist cloud systems, and
therefore also regions of radiation minima. The lowest radiation
temperature of this part of the pressure field, averaged for 20
situations, is —52° (with a range of —40 to —60°). The area occupied
by the cloud system is on the average $8 \times 10^5 \, km^2$.

In contrast, to the east of the ridge and up to the axis of the
trough there is usually clear sky, and this is confirmed by the order
of magnitude of the radiation temperatures: in the east part of the

blocking ridge, as an average for 20 situations it is 0°, ranging from + 5 to —10°. The lowest temperatures, typical for the periphery of the ridge and determined by the isotherm surrounding the ridge, are from —15 to —20°.

It is noteworthy that in the cloud-free part of the ridge the radiation temperature isotherms repeat on general lines the configuration of the 500 mb isohypse. Usually the 584 and 580 decameter isohypses coincide with the 0 and + 10° isotherms, whereas the 560 decameter isohypse coincides with the —20 and —30° isotherms. As a result, the position of these isotherms on radiation temperature maps gives an approximate picture of the position of the ridge.

The change in position of the radiation minima related to the position of the depression centers can serve as a characteristic of the evolution of depressions. This change can be traced by the position of the surface lows and the radiation minima in Figure 2. For an appraisal of the stage of development of the depression the lowest pressure is shown on Figure 2 by double symbols.

The cycle of displacement of the radiation minimum related to the surface low center can be traced in more detail for depression D_b in January.

In the earlier stages (13 January) the radiation minimum lay near the cyclone center. Further on, it began to be displaced relative to the surface center, so to speak circling it in an anticlockwise direction. During deepening (17 and 18 January) it moved to the southeast quadrant, and in the filling stage it was observed in the north quadrant. Along general lines a similar displacement of the radiation minima was also observed for lows D_c, D_a and D_d.

The features of relative distribution of these centers in different development stages of the depression, as described above, are also confirmed by some examples for February (D_a, D_d, D_e on Figure 2).

Some differences are noted in the displacement of the radiation minimum of depression D_d in January. This minimum remained in the southeast quadrant also in the filling stage of the low (Figure 2). The displacement of this radiation minimum, being connected with cloudiness, was in this case held up directly by the adjacent ridge to the east whose northern part was slightly curving westward (Figure 2).

A somewhat similar situation arose also in February with weak lows D_b and D_c. In their filling stage on 8, 9 and 10 February the radiation minima, which lay directly near the cyclonic centers, also did not reach the northeast quadrant. The displacement of the cloud systems was in this case too hindered by the vertex of a ridge strongly bent westward.

Thus, we may summarize the following features of the relative positions of radiation and pressure centers during the latter's evolution. In the development stage of the depression the radiation minimum lies near to it. With the development of the depression it begins to move southeast and then to the northeast quadrant. When the radiation minimum is in the southeast quadrant the depression is at its maximum depth one to two days before or after it, while when the minimum is in the northeast quadrant and on the northern periphery of the depression the latter is in its filling stage. The velocity of movement of the radiation minimum is determined by the particular features of the upper-level pressure field.

To determine the possibility of estimating the depth of pressure centers from the radiation temperature field, we studied the diurnal change of extreme radiation temperatures in parallel with the change in intensity of the pressure centers.

In cloudless weather we observed at the eastern periphery of a blocking ridge in 17 cases out of 20 the same signs in the change of radiation temperature corresponding to the northern extremity of the 556 decameter isohypse. This value characterizes the degree of northward extension of the ridge, and therefore also of its intensity.

In cyclonic weather the sign of change of the radiation temperature corresponded in 7 out of 18 cases to the sign of change of the pressure in the depression center, while in 5 cases it was not so. In the remaining 6 cases, when there was no pressure change in the depression center, the change in radiation temperature was insignificant (up to 5°) or zero.

With insignificant pressure changes in the depression center (from 0 to 5 mb/day) no clear relation was observed between the sign of pressure change and that of radiation temperature. In the only case (of 18 investigated) when there was a sharp pressure fall (25 mb/day), there was also a sharp change in the lowest radiation temperature (19°).

For determining the stages of depression development we also used the orientation of the radiation minima, which defined the change in the orientation of fronts in the course of depression evolution. During the growth stage the radiation minimum was oriented in a latitudinal direction, corresponding to the orientation of the fronts. This was clear from the few examples at our disposal.

The radiation minima corresponding to a deepening low were oriented from northeast to southwest or meridionally. A tilted orientation of the radiation minimum is typical for a cold front, while a meridional direction indicates an occlusion, and often still appears up to the time when the depression reaches its lowest central value.

The orientation of the radiation minima related to a filling depression is usually meridional. A number of circular minima were related to the last stage of the depression, when cloudiness was maintained only to the north of the depression.

5. CLOUD FIELD ACCOMPANYING LATITUDINAL CYCLONIC TRANSFORMATIONS

As indicated before, in February a stable ridge lay along the west coast of North America, extending northward during the first half of February (Table 2). The occurrence on 15 February of a latitudinal cyclonic transformation intersected this narrow ridge and as a result its northerly boundary, formed by the 556 decameter isohypse, shifted from 70 to 40°N, while in the north a cut-off anticyclonic center formed (Figure 2). The beginning of this transformation may be attributed to the formation on 6 February of an upper low of 564 decameters to the west of the blocking ridge. This center extended up to 200 mb, but did not appear on the surface and at 850 mb (Figure 2). From 6 to 13 February it maintained its position in the San Francisco region with hardly any change, displacing itself to the east and deforming the southern part of the blocking ridge, so that the ridge tilted to the northeast of this upper low. On 12 February there appeared, instead of the displaced ridge, a new ridge to the south at 140° which merged with the previously existing anticyclone to the north (Figure 2).

On 11 February a new upper level cyclonic center (540 decameters) appeared to the west of this ridge, moved east, intersecting the ridge and advancing into the continent (Table 1, Figure 2b).

The latitudinal transformation finally ended on 14 February with the formation of yet another cyclonic center of 548 decameters in the west which, together with the preceding center of 540 decameters, formed a latitudinally oriented cyclonic belt, making the process similar to a zonal one.

A high-pressure region to the north, bounded by the 556 decameter isohypse, was maintained on 16 and 17 February, but subsequently broke up.

The causes of appearance of three cyclonic centers (on 6, 11, and 14 February) against the background of high pressure, and broken up by the influence of the blocking high, can be seen in the redistribution of heat flux in the atmosphere.

Monin /4, 5/ has repeatedly indicated the role of cloudiness as "an effective regulator in processes of redistribution of the constant

flux of solar energy in the nonuniform time and space distribution
of heat flux in the atmosphere," which in turn can change the form
of the pressure field.

When examining the regulating influence of clouds it is important
to treat separately those regions in which cloudiness is a constant
feature, and therefore there may exist some considerable anomalies
in the distribution of heat flux. In addition, the above conclusions
concerning the predominance of cloudiness in well defined parts of
the upper-level pressure field may be useful: to the east of the
axes of troughs and to the west of the axes of ridges, where cloud
systems emanating from the west are held up and accumulate,
favorable conditions are created for the redistribution of the heat
flux in the atmosphere.

A zonal flow is less favorable for the maintenance of cloud fields
in the same region. In these conditions cyclonic systems move
with a velocity of 35 — 40°/day, appearing only temporarily over a
certain area. Therefore, under zonal flow, the influence of clouds
on the redistribution of heat flux is not so considerable, and one
would rather expect a stronger influence of the underlying surface
on the change in the pressure field.

During two months we noted four cases of the appearance of
depressions (17 January and 6, 11 and 14 February) in regions with
stationary cloudiness to the west of an anticyclone. This can be
seen on specially constructed maps of frequency of cloudiness for
four days preceding the appearance of each of these lows (Figure 3).
The positions of the depressions appearing on the fifth day and the
position of the ridges blocking the advance of cloud systems were
also plotted on these maps.

The upper-level low appearing on 6 February was accompanied
by an isolated cold pool. The low extended to the 200-mb level, but
did not appear on the surface. On 11 February this depression
was above the old filling surface low where the cloud system of
the occlusion was still maintained, and which was held up by the latitudinally curved blocking ridge (Figure 3). On February, over the
region of greatest frequency of cloudiness, separate cyclonic centers
split off from the main upper low.

The coincidence of the appearance of upper low centers in the
same place as the stationary cloud system to the west of the blocking ridge confirms the concepts of Monin about the possible influence of cloudiness as a feedback mechanism in the change of
pressure fields.

However, this same influencing mechanism of the cloud cover
can develop only after a redistribution of the heat flux in certain
synoptic situations.

The above-examined upper cyclonic centers, causing latitudinal cyclonic transformations, are usually associated with cloud systems of special configuration and orientation, being different from the cloudiness of the intertropical zone and the cloudiness forming on the southern periphery of the subtropical anticyclone /1/. These very large cloud systems are arranged in a trailing band, broadening considerably toward the south. They are oriented almost meridionally between 40 and 10°N. The physical mechanism of the formation of very widespread and persistent cloudiness in such conditions can be attributed to upper-level convergence connected with the approach of an upper-level low to a blocking ridge.

6. IMPROVEMENT OF FRONTAL ANALYSIS BY SATELLITE DATA

Satellite information is useful in frontal analysis, especially over oceans. It is especially valuable for the study of evolution of synoptic processes, i. e., it can be used to achieve continuity in synoptic analysis. Here, the cloud field viewed by satellites can either confirm the analysis or serve as a prerequisite for greater accuracy, or as its refutation.

An examination of operational synoptic charts of January and February 1968, analyzed without detailed use of satellite data, showed that in most cases (74 out of 121) a pronounced cloud band corresponded to fronts (Table 3). In operational synoptic analysis, good agreement of fronts with cloud bands on the basis of satellite data was noted in 28 cases for cold fronts, in 25 cases for warm fronts, and in 21 cases for occlusions.

A survey of the cases of discrepancy between operational synoptic analysis and satellite data yielded the following typical cases of inaccuracy in the placing of fronts.

The occlusion process in fast moving depressions over the ocean occurs in a shorter period than that displayed on synoptic maps by frontal analysis.

Many depressions which, on synoptic charts, conserve a broad warm sector are already occluded, judging from the cloudiness. The mean value for the narrowing of the warm sector according to 22 examples with an accelerated occlusion is 40° of the arc of latitude taken around the point of occlusion. A much earlier occlusion occurred in the examined examples under the influence of a blocking ridge, which held up the advance of the warm front with simultaneous acceleration of the cold front over the ocean surface (Table 3).

TABLE 3. Comparison of frontal positions according to synoptic analysis and on the basis of satellite data

Agreement			Discrepancy				
situation	number of cases	particulars of situation	difference in position in degrees of arc of the latitude circle			number of cases	
			mean	maximum	minimum		
Cold front ...	28	Accelerated advance of cold front after earlier occlusion	4	8	2	22	
Warm front ..	25	Large curvature of occluded front	3	4	2	14	
Occlusion ...	21	Displacement of cold front into subtropical latitudes	4	6	2	11	

A decrease in the velocity of motion of cold fronts can be explained by the straight parts of occlusions, appearing on synoptic maps, whereas according to the cloudiness occluded fronts occur in the form of strongly bent spirals /3/. Thus, satellite information leads to the conclusion that Bergeron's ideas about backbent occlusions are near the truth.

Another discrepancy between synoptic analysis and nephanalyses, confirmed by 11 examples for the east Pacific Ocean, relates to the more northerly position of cold fronts on synoptic maps (Table 3). This essentially applies also to the decrease in velocity of motion of cold fronts. These divergences may be connected with the traditional opinion that cold fronts cannot penetrate into subtropical latitudes. In effect, ideas to the contrary have recently been voiced /2/. In some investigations of such examples the fronts entered on the synoptic maps on the basis of surface temperatures are on the whole correct. However, a complicating factor is the frontal cloudiness which, at night, favors a temperature rise at the surface. This causes the main temperature contrasts and consequently also the front to be displaced to the north on synoptic maps.

Finally, the last typical improvement in frontal analysis is that, according to satellite data, the cloud system of an occlusion is maintained longer than the occluded front itself on synoptic maps (11, 18, 19 and 20 February). The discrepancy of one to two days is quite understandable, since with an old occlusion the boundary of frontal division at the surface becomes blurred, while the

cloudiness still persists.   This indicates some inertia of the cloudiness.   In this sense the satellite measurements give real supplementary information, facilitating the continuity of consecutive
analyses.

BIBLIOGRAPHY

1. Belov,P.N., Yu.V.Kurilova, and I.R. Egorova. Osobennosti oblachnykh polei v
   tropikakh po radiatsionnym izmereniyam so sputnikov (Features of Cloud Fields in the
   Tropics on the Basis of Satellite Radiation Measurements). — Trudy Gidromettsentra
   SSSR, No. 50. 1969.
2. Kaplina,L.I. and L.S.Minina. Kharakteristika oblachnosti kholodnykh frontov v
   nizkikh shirotakh Tikhogo okeana (Characteristics of Cloudiness of Cold Fronts in Low
   Latitudes of the Pacific Ocean). — Trudy Gidromettsentra SSSR, No.41. 1969.
3. Leonov,N.G.,T.P.Popova,A.M.Tsar'kova, and E.F.Shumanskaya. Sinopti-
   cheskie usloviya sushchestvovaniya oblachnykh vikhrei (Synoptic Conditions in the
   Formation of Cloud Vortices). — Trudy Gidromettsentra SSSR, No.11. 1967.
4. Monin,A.S. O fizicheskom mekhanizme dolgosrochnykh izmenenii pogody (On the
   Physical Mechanism of Long-Term Weather Changes). — Meteorologiya i Gidrologiya,
   No.8. 1963.
5. Monin,A.S. Prognoz pogody kak zadacha fiziki (Weather Prognosis as a Physical Problem).—
   In Sbornik: "Nauka i chelovechestvo." Moskva. 1964.

UDC 551.507.362.2:551.524.7

VARIABILITY OF EIGENVECTORS AND EIGENVALUES OF CORRELATION MATRICES FOR VERTICAL TEMPERATURE PROFILES

V. G. Boldyrev

This paper examines the latitudinal dependence of the second and third eigenvectors of auto-correlation matrices for vertical temperature profiles.

In computing eigenvalues and eigenvectors, previously obtained empirical autocorrelation matrices, relating to meridional cross sections over European USSR and the Atlantic Ocean, are used. It is found that the universality of the second and third eigenvectors can be admitted only in certain limits. This statement relates also to the eigenvectors of normalized correlation matrices, especially to vectors higher than second.

At the present time a number of problems arise whose solution requires data on the statistical structure of the vertical temperature profile in the atmosphere. One such problem is the reconstruction of the vertical temperature profile from data on the spectral distribution of emitted radiation obtained from earth satellites /3, 6/. Some methods of solving these problems and also the determination of the information value of the measurements /2, 6/ require a knowledge of the eigenvectors and eigenvalues of the autocorrelation matrices of the vertical temperature profile.

It follows from /8/ that these eigenvectors, forming an orthonormal basis, allow an optimum approximation to the vertical temperature profile, treated as a random function.

In /6/, this orthonormal basis was first applied to solve the practical problem of determining the vertical temperature distribution from the spectrum of emitted infrared radiation. The authors show the advantage of using the above-mentioned basis systems.

Subsequently, the results of computing eigenvalues and eigenvectors by studying the basic properties of these values were investigated /4, 5/. In particular, the dependence of empirical eigenvectors on the time of year, time of day and type of terrain was studied /5/. The statistical stability of the first eigenvector was established, and conclusions were drawn regarding the universal statistical applicability of the system of eigenvectors.

The dependence of the first eigenvector on latitude was studied for temperature /4/. This dependence turned out to be rather weak.

At the same time it was revealed /4, 5/ that the eigenvectors with higher numbers (e. g., second and third) already cannot be regarded as independent of geographical position, season, etc. However, these vectors are required in the approximation of the vertical temperature profile, since their role in the total variance of the expansion coefficients is quite considerable /4/.

In this paper we examine in more detail the dependence of vectors with higher numbers than one on latitude and type of terrain.

Material for the computation of eigenvectors and eigenvalues consisted of autocorrelation matrices for vertical temperature profiles, evaluated for two meridional cross sections: over the North Atlantic and European USSR for winter and summer /1/. As a rule all correlation matrices were calculated for six atmospheric levels: 1000, 850, 700, 500, 300, 200 mb. The number of thus realized vertical temperature profiles used in the calculations for most radiosonde points was 300 — 400, and for no station were less than 200 used. According to estimates made in /7/ this number of actual realizations provided in every case sufficient accuracy of computation for the second and third eigenvectors.

Turning to the results of computations, let us examine first the latitudinal dependence of the role played by the different eigenvectors in the total variance. Table 1 shows this contribution for the first vector, then the total for the first and second and for all three vectors. It is easily seen that in a given latitude zone (approximately from the Black Sea to Arkhangelsk) the role of the total variance described by some combination of eigenvectors hardly varies either longitudinally or with the season, whatever the change occurring in the absolute variance. One can notice only a weak tendency to a southward decrease in the role played by the first vector on account of some increase in the role of the second and third vectors (and also in the contribution of subsequent vectors). Bearing this in mind, it would be interesting to examine in particular the eigenvalues and eigenvectors of the temperature correlation matrices referring to tropical latitudes.

Although we shall not develop here the data relating to the North Atlantic it is noteworthy that, in this region too, there was a tendency toward a decreasing contribution of the first eigenvector in southerly latitudes. This tendency, however, here becomes apparent not from 45 — 50°N but from 35 — 40°N.

We now examine the latitudinal dependence of these eigenvectors (the second and third).

TABLE 1. Latitudinal dependence of the residual variance $K_n = \sum_{n=1}^{n} \lambda_n \Big/ \sum_{n=1}^{N} \lambda_n$, European USSR (%)

n	Latitude, degrees				
	45	50	55	60	65
1	0.41	0.59	0.65	0.63	0.64
	0.51	0.59	0.66	0.68	0.58
2	0.66	0.78	0.81	0.79	0.82
	0.67	0.81	0.83	0.79	0.81
3	0.87	0.93	0.92	0.89	0.92
	0.81	0.92	0.90	0.88	0.92
$\sum_{n=1}^{N} \lambda_n$	112.11	110.18	106.25	113.78	128 60
	121.80	151.23	230.52	240.47	214.77

Note. The first row gives values of K_n and $\sum_{n=1}^{N} \lambda_n$ for summer, and the second row for winter.

Figure 1 gives some idea of the nature of this dependence in winter. The figure shows meridional cross sections of the second and third eigenvectors for European USSR.

TABLE 2. Cosines of angles between eigenvectors of nonnormalized correlation matrices. North Atlantic, January

		ψ_1	ψ_2	ψ_3	ψ_4	ψ_5	ψ_6
φ_1	1	0.96	—0.18	—0.16	0.36	—0.01	—0.05
	2	0.94	—0.09	0.32	—0.10	0.13	0.07
	3	0.82	—0.28	—0.48	0.41	0.10	—0.13
φ_2	1	0.22	0.96	0.05	—0.34	0.14	0.03
	2	0.30	0.97	0.05	0.14	—0.09	—0.04
	3	0.49	0.86	0.01	—0.18	—0.01	0.06
φ_3	1	—0.12	0.13	0.97	0.22	—0.28	—0.02
	2	—0.01	0 12	—0.33	—0.89	—0.17	—0.08
	3	—0.08	—0.06	0.13	—0.51	0.31	0.20
φ_4	1	0.19	0.15	0.26	0.44	—0.40	—0.07
	2	0.09	0.04	0.75	—0.42	0.27	0.15
	3	—0.08	0.06	—0.93	0.10	0.27	—0.19
φ_5	1	0.03	—0.06	0.02	0.32	0.87	—0.27
	2	—0.22	0.01	—0.30	—0.08	0.89	0.27
	3	—0.17	—0.04	0.02	0.05	0.79	—0.49
φ_6	1	—0.03	0.04	0.01	—0.19	—0.20	0.96
	2	0.40	0.67	—0.29	0.31	0.31	0.96
	3	0.64	0.57	0.37	0.28	—0.47	0.88

Note. Here and in Table 3: 1 — φ (ship C), ψ (ship D); 2 — φ (ship E), ψ (ship C); 3 — φ (ship E), ψ (ship D).

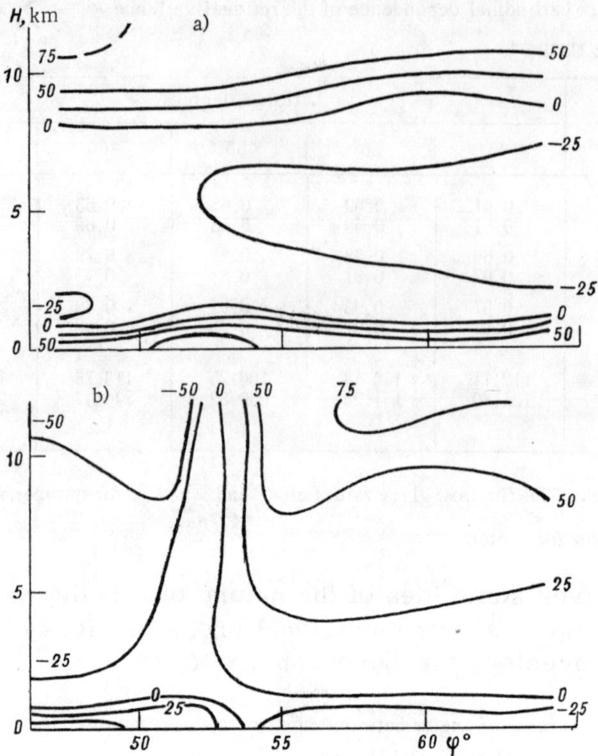

FIGURE 1. Meridional cross sections of the second (a) and
third (b) eigenvectors of temperature correlation matrices for
European USSR in winter.

In winter, meteorological conditions over all European USSR are
generally uniform.   The meridional cross section of the second
eigenvector does not reveal any considerable dependence on latitude,
but the cross section of the third eigenvector indicates the existence of two zones with a different behavior of this vector, the limit
between these regions lying at approximately $52-54°N$.

The cross sections of the second and third eigenvectors, obtained
for the North Atlantic, differ considerably from the cross sections
of the same vectors for European USSR.   This difference is especially noticeable near the earth's surface (second vector).   The
latitudinal dependence of the vectors in this oceanic cross section
is especially marked south of 45°N.

Thus, the universality of the second and third eigenvectors can
be assumed only within certain limits.   It is interesting to clarify
to what extent a similar assumption can influence the quality of

solution of some concrete problem (such as the "inverse" problem of satellite meteorology).

TABLE 3. Cosines of angles between eigenvectors of normalized matrices. North Atlantic, January

		ψ_1	ψ_2	ψ_3	ψ_4	ψ_5	ψ_6
φ_1	1	1.00	0.05	0.02	—0.14	—0.01	0.00
	2	0.96	0.21	—0.17	—0.09	—0.05	—0.02
	3	0.98	0.16	—0.16	—0.04	—0.02	—0.04
φ_2	1	—0.05	0.95	0.27	—0.10	—0.15	—0.01
	2	0.25	0.90	—0.24	—0.29	—0.11	0.00
	3	—0.20	0.77	—0.50	—0.39	0.34	0.09
φ_3	1	0.01	0.31	0.89	—0.27	—0.10	—0.08
	2	0.07	0.32	0.81	0.35	0.02	—0.08
	3	0.01	0.63	0.48	0.49	0.10	—0.02
φ_4	1	0.04	0.02	0.19	0.82	—0.42	0.36
	2	—0.03	0.16	0.43	0.66	—0.32	—0.18
	3	—0.15	0.00	—0.66	0.69	0.14	—0.27
φ_5	1	0.02	0.09	0.22	0.46	0.74	—0.43
	2	0.00	0.18	0.27	0.23	0.95	—0.06
	3	0.33	—0.05	—0.03	—0.02	0.84	0.53
φ_6	1	0.01	—0.05	—0.10	0.16	—0.36	0.82
	2	—0.01	—0.06	0.02	—0.20	0.01	0.97
	3	0.00	—0.02	—0.13	0.35	—0.54	0.79

The difference in the eigenvectors is of course connected with the difference in the corresponding correlation matrices. By normalizing the correlation matrices we can obtain great similarity between the normalized matrices (of the correlation coefficients) and the covariant matrices, since in this way the heterogeneity of the variance is accounted for. It remains to point out that at the same time the latitudinal dependence of the eigenvectors of the normalized matrices remains.

To illustrate this statement, Table 2 presents the cosines of the angles between the eigenvectors of nonnormalized correlation matrices relating to cross sections over the North Atlantic, and Table 3 gives similar values for normalized correlation matrices. The data of these tables show that the systems of eigenvectors are especially different for the third and fourth vectors. The fifth and sixth vectors (which constitute together 5 to 10% of the total variance) can already be compared with some "noise," which is approximately uniform for all latitudes.

The third and fourth vectors (which account in the examined cases for up to 13 and 9% of the total variance, respectively) undoubtedly play a more important role. The expansion terms

corresponding to these vectors can hardly be neglected.   Therefore
it is necessary to account for the dependence of the corresponding
expansion coefficients on the actual conditions.

BIBLIOGRAPHY

1. Boldyrev, V. G. and V. M. Oleshev. O statisticheskoi strukture vertikal'nykh profilei
    temperatury i vlazhnosti (Statistical Structure of Vertical Temperature and Moisture
    Profiles). — Trudy MMTs, No. 11. 1966.
2. Kozlov, V. P. O vosstanovlenii vysotnogo profilya temperatury po spektru ukhodyashchego
    izlucheniya (Reconstruction of Upper-Level Temperature Profiles According to the
    Spectrum of Emitted Radiation). — Izvestiya AN SSSR, Seriya Fizika Atmosfery i
    Okeana, Vol. 2, No. 2. 1966.
3. Kondrat'ev, K. Ya. Meteorologicheskie sputniki (Meteorological Satellites). Leningrad,
    Gidrometeoizdat. 1964.
4. Koprova, L. I. and V. G. Boldyrev. O statisticheskikh kharakteristikakh vertikal'noi
    struktury polei temperatury i vlazhnosti do bol'shikh vysot (Statistical Characteristics of
    the Vertical Structure of Temperature and Moisture Profiles up to Great Heights). —
    Izvestiya AN SSSR, Seriya Fizika Atmosfery i Okeana, Vol. 6, No. 2. 1970.
5. Koprova, L. I. and M. S. Malkevich. Ob empiricheskikh ortogonal'nykh funktsiyakh
    dlya optimal'noi parametrizatsii profilei temperatury i vlazhnosti (Empirical Orthogonal
    Functions for Optimum Parametrization of Temperature and Moisture Profiles). —
    Izvestiya AN SSSR, Seriya Fizika Atmosfery i Okeana, Vol. 1, No. 1. 1963.
6. Malkevich, M. S. and V. I. Tatarskii. Opredelenie temperatury i vlazhnosti zemnoi
    atmosfery po izmereniyam izlucheniya Zemli so sputnikov (Determination of Tempera-
    ture and Moisture of Earth's Atmosphere from the Earth's Radiation as Measured from
    Satellites). — In Sbornik: "Issledovaniya kosmicheskogo prostranstva." Moskva,
    Izdatel'stvo "Nauka." 1965.
7. Meshcherskaya, A. V. et al. Estestvennye sostavlyayushchie meteorologicheskikh polei
    (Natural Components of Meteorological Fields). Leningrad, Gidrometeoizdat. 1970.
8. Obukhov, A. M. O statisticheski ortogonal'nykh razlozheniyakh empiricheskikh funktsii
    (Statistical Orthogonal Expansion of Empirical Functions). — Izvestiya AN SSSR,
    Seriya Geofizicheskaya, No. 3. 1960.

UDC 551.507.362.2:551.521.14

STATISTICAL STRUCTURE OF THE BRIGHTNESS FIELD OF REFLECTED RADIATION IN THE 0.6 −0.8 MICRON SPECTRAL INTERVAL

V. G. Boldyrev, V. I. Tulupov

This paper examines the results of statistical processing of continuously recorded measurements of radiation reflected from the earth's surface and cloud tops in the 0.6 − 0.8 micron range of the spectrum. These measurements were made by an instrument installed on the Cosmos 121 satellite. The computed characteristics of the space structure of the radiation field (correlation functions and spectral density) indicated an absence of pronounced oscillations in the mesoscale range of the spectrum, in the total for all investigated cases as well as for individual cases relating to actual meteorological conditions.

At present we have conducted some quantitative statistical investigations of the earth's radiation field in different parts of the spectrum /1 −3/. These studies use, in particular, measurements of short wave radiation obtained by different satellites whose spatial resolution at the nadir was usually several tens of kilometers (30 − 70 km). This hampered the computation and analysis of properties of the space structure of the radiation field in actual synoptic or geographical formations, since there were a few independent measurements relating to the same formation. Second, the averaging, introduced by the carried instruments, limited the high frequency regions of the energy spectra and did not allow more detailed study of the interval of the mesoscale minimum.

It was noted /1/ that a sharp decrease in the spectral density of reflected radiation takes place, beginning from long waves of the order of 700 − 1000 km. In addition, sliding averages were taken over intervals of about 5000 km. In this way we filtered out practically only the spatial trend caused by different heights of the sun at different latitudes.

There is also some interest in producing further filtration in the low frequency band to better understand the oscillations in the reflected long wave radiation of less than 700 − 1000 km.

The present work is an experimental analysis of this kind based on processed observations of radiation reflected from the earth's surface and cloud tops, obtained from the Cosmos 121 satellite.

Since the space resolution of the instrument was quite high, it was
also possible to study the structure of different clouds and geograp
ical formations.

The intensity of reflected and scattered radiation in the 0.6 — 0.8
interval on Cosmos 121 was measured in the direction of the nadir.
The angle of the field of view of the photoelectric photometer used
was 1°, with a resolution power of 4 — 6 km. The instrument worke
continuously over the daylight side of the earth during several days
The information was recorded on photographic film by an instrument installed on the satellite.

The areas of investigated surface were filmed by a three-lense
objective with a focal length of 41 mm and optical power 1:2 on a
viewing slot and through a two-lense condensor, and then projected
on a sensitive element.

The sensor of the photometer consisted of a photoresistor of
cadmium selenide. The long wave range of the working interval
was $0.6 - 0.8\mu$ and was determined by the spectral sensitivity of
the receiver; the short wave limitation was given by a detachable
light filter, admitting long wave radiation exceeding 0.6μ. The
maximum sensitivity of the photometer lay at 0.67μ.

The instrument was calibrated by sunlight reflected from a
screen covered by magnesium oxide. There is some nonlinearity
on the calibration curve caused by nonlinearity of the photoresis-
tance with low energies, or by the electrical layout for broadening
the dynamic range of the instrument for large energies. However,
since the main measurements were in the limits where the calibra
tion curve can be considered linear, neglecting this nonlinearity
will not considerably distort results.

The measurements were processed for six daytime passes
(20 June 1966) over Central Asia and European USSR, over West
Africa, western Europe and the East Atlantic. Television pictures
of cloudiness obtained on this day and at approximately the same
time give evidence of the great variety of cloud forms over the
above-mentioned regions.

Radiation data (in relative units) were taken with a recording
instrument so that the distance between separate readings corresponded to a distance on the surface of 2 km between separate
measurements. This ensured approximately two measurements
for one element of resolution. The length of the measured section
was about 10,000 km. The data obtained in this way were processe
separately for each pass and then the computed statistical properties (when necessary) were averaged over all the measured strips.
Processing during the computation also included some statistical
properties (mean, variance, correlation and structural functions)

for several sections of 500 — 1000 km length, in addition to the above-mentioned properties for all cases. In the last case we applied sliding means of different scales, but in every case the averaging interval did not exceed 1000 km. By this method we filtered large-scale synoptic oscillations but, it seems to us, not completely so, since the frequency characteristic of this type of filter is far from ideal.

The sliding means were then found, and for the obtained deviations we computed correlation functions

$$K(\tau) = \overline{x(t)\,x(t+\tau)}, \tag{1}$$

where x is the deviation from the sliding mean and τ the space displacement.

We also computed structural functions

$$B(\tau) = \overline{[x(t) - x(t+\tau)]^2} \tag{2}$$

and compared correlation and structural functions to estimate the steadiness of the analyzed series.

Thus the mean of the correlation functions of the reflected radiation was derived on the basis of all processed passes (Figure 1a). This function has an exponential cosine character. The correlation function passes through zero already at distances of about 200 km. Comparison of this correlation function with those given in /1/ (where the correlation function was obtained with a sliding mean interval of about 5000 km) illustrates very significant weakening of the spatial correlation.

The function computed in /1/ passed through zero at distances of approximately 700 km, which is comparable to the corresponding distance observed with the correlation function for cloudiness. According to /4/ the latter correlation function, computed from surface observations, passes through zero at about 1000 km.

The fact that the damping of the correlation function shown in Figure 1a occurs more rapidly gives evidence of a lack of regularity (in the statistical sense) of the variation in the values of cloud brightness in synoptic-scale systems. Of course such a situation cannot occur in conformity with this or that actual cloud formation (e. g., a bank of cumulus clouds), but on the average one must remember that spatial relations of brightness (and therefore cloudiness) on a scale less than 1000 km are very much weaker than those occurring on a synoptic scale.

We now estimate the contribution made by a variation in some frequency within the total spatial variation in the intensity of reflected radiation.

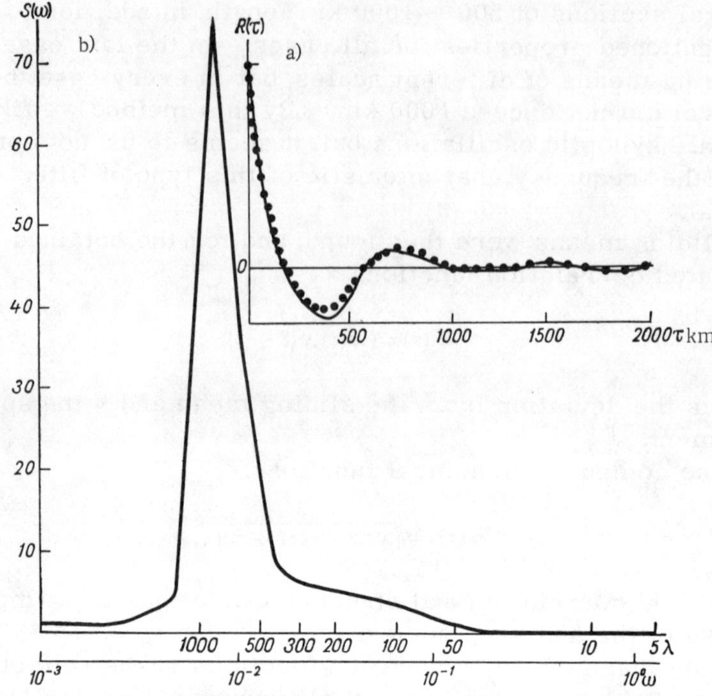

FIGURE 1. Mean correlation function for reflected radiation (a) and
mean spectral density of reflected radiation (b).

For this purpose we calculated the spectral density of this quantity with the aid of the mean correlation function. The correlation
function was first approximated by the expression

$$R(\tau) = 0.8e^{-\alpha\tau}\cos\beta\tau - e^{-\gamma_1\tau} + e^{-\gamma_2\tau} + 0.2e^{-\tau}, \qquad (3)$$

where $\alpha = 0.003$, $\beta = 0.0075$, and the values of γ_1 and γ_2 are 0.015
and 0.036, respectively; the value of τ is expressed in km. The
approximate values of the correlation function are indicated on
Figure 1a by circles.

The spectral density is hence given by the following expression:

$$S(\omega) = \frac{2}{\pi}\left[0.8\alpha\,\frac{\omega^2 + \alpha^2 + \beta^2}{\omega^4 + 2(\alpha^2 - \beta^2)\omega^2 + (\alpha^2 + \beta^2)^2} - \right.$$

$$\left. - \frac{\gamma_1}{\gamma_1^2 + \omega^2} + \frac{\gamma_2}{\gamma_2^2 + \omega^2} + \frac{0.2}{1 + \omega^2}\right]. \qquad (4)$$

The value of S, as calculated according to formula (4), is shown in Figure 1b, where frequency ω and wavelength λ are marked on the abscissa.

The obtained spectral density is naturally very similar to that given in /1/. It is, however, clear that in spite of the filtration of low frequencies of oscillation ($\lambda > 1000$ km) the spectral density does not reveal any significant features for long waves corresponding to mesoscale processes. As before, we note the sharp decrease in spectral density with decrease in wavelength; this again confirms the presence of mesoscale minima in the spectrum of reflected radiation and therefore the spectrum of the cloud field.

For a more detailed investigation into the mesoscale range of the spectrum, data on reflected radiation during several passes were processed by sliding averages over various intervals. Figure 2a shows correlation functions corresponding to the same pass with sliding averages over intervals of 1000, 500, 200, 100 and 50 km. It is easily seen that with oscillations on a scale less than 50 km we practically arrive at a "white noise" — an uncorrelated deviation from the sliding average. The residual variance of this noise comprises rather a large proportion (about 20%) of the variance of all mesoscale oscillations.

Thus, one can conclude that from the standpoint of investigating large-scale energetic processes, averaging over an area of about 1000 km is sufficient, and a more detailed study does not yield any important information. Nevertheless, there are variations on a scale of less than 5 km (in our case this is the instrument's resolution at the nadir) which contribute considerably to the total variation of the process.

Data on variations in brightness and on the spatial structure of the brightness field are important in the analysis of TV and infrared cloud photographs. For this purpose a detailed measurement of brightness is required with a resolution of less than 5 km.

The above discussion relates to some mean property obtained by processing meteorologically very diverse cases.

It is also of interest to examine the statistical properties of certain cloud systems and types of underlying surface, such as desert or ocean in cloudless conditions. We chose some sections of the recorded measurements of such surfaces and computed the spatial correlation functions for each section. The selected sections were no longer than 1000 km in order to avoid possible spatial trends. If such a trend was nevertheless discerned, long wave oscillations were filtered. For convenience in comparing different correlation functions, all long wave oscillations of more than 400 km were filtered out.

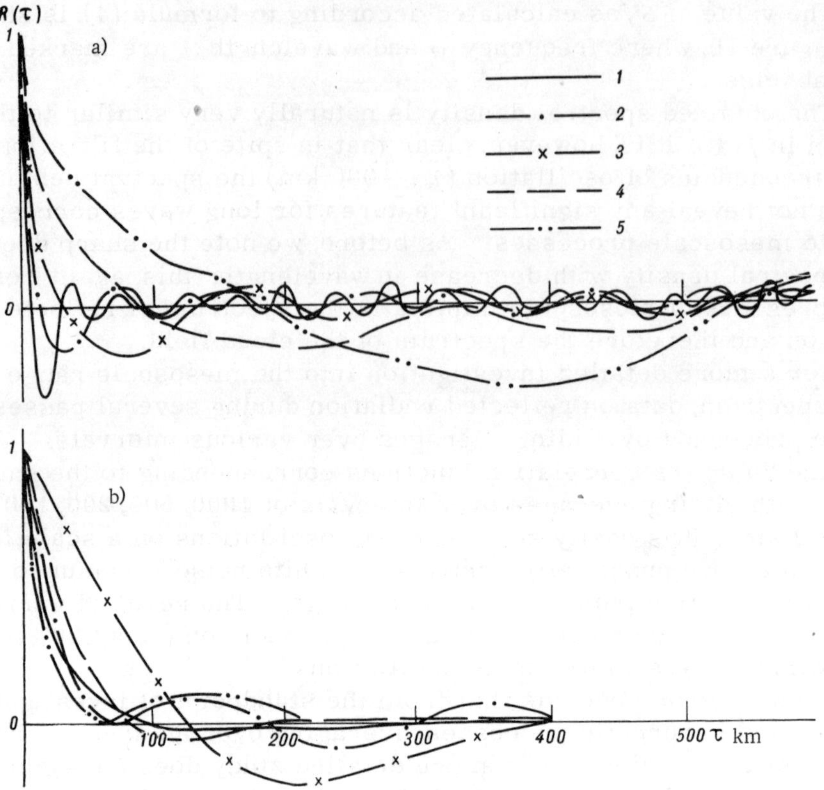

FIGURE 2.   Correlation functions for sliding averages over different intervals (a) and correlation functions for several situations (b):

1 — 50 km; 2 — 100 km; 3 — 200 km; 4 — 500 km; 5 — 1000 km.

In this way one can assess the nature of mesoscale perturbations inherent in selected cases.

It would be premature to generalize about some types of clouds or underlying surface on the basis of an insufficient number of cases. The aim here is simply to present some examples of correlation functions for different situations.

Figure 2b presents correlation functions for several situations. Curve 1 relates to cirrus clouds and nimbus forms such as As and Ns, found in a frontal zone extending from Iceland through north Scandinavia to Novaya Zemlya.   The orbital plane did not intersect the frontal zone, but passed along it in the examined region.   The correlation functions have no oscillating components, since the long-wave oscillations were almost indiscernible owing to the very character of the case.

Curve 2 relates to the Sahara region and lies to the west of the Tassili plateau. Surface observations showed that the region was cloudless. On the TV pictures some nonuniformity in brightness could be discerned, probably associated with variations in the reflective properties of the surface. In the correlation functions a rather regular periodical oscillation with a period of 200 km could be noted. It is doubtful whether this periodicity was connected with the existence of clouds (not visible in the TV photos), or simply reflects some property of the terrain.

The correlation function illustrated in curve 3 relates to a region of thick clouds, connected with a barometric depression lying south of Iceland. Here one can notice a well-marked exponential cosine character of the correlation function, corresponding to the appearance of a long wave oscillation of 400 — 500 km which it was not possible to eliminate altogether by filtration of the empirical correlation function. Instead one can see that, on the whole, the statistical structure of the intensity of reflected radiation is not considerably different from that of the case examined before.

Curve 4 is a correlation function belonging to a cumulus cloud system over the Atlantic, a little west of the Bay of Biscay. Here, the cumulus clouds were not continuous, and on TV photographs they stood out clearly against the background of dark ocean. At first glance one might expect the occurrence of some periodic mesoscale oscillations in the intensity of radiation. However, it is easily seen that this is not the case; mesoscale variations occur in an unordered manner and do not show any periodicity.

Finally, the correlation function illustrated by curve 5 relates to a region with massive continuous cloudiness of stratiform clouds and nimbostratus in the Atlantic around 55°N and 25°W. Here, as before, mesoscale oscillations are absent; there are only oscillations with a period near 250 km which may reflect the presence of breaks in cloudiness, i. e., some unsteadiness in the examined phenomena.

The following conclusions may be drawn:

1) regular mesoscale variations in the field of reflected radiation and consequently the field of cloudiness are apparently rather frequent;

2) the spatial correlation functions for the intensity of reflected radiation, relating to the mesoscale range of the spectrum (wavelength 5 — 400 km), are only weakly dependent on the type of object reflecting radiation; hence it is doubtful whether one can use them as quantitative indicators of these objects;

3) as regards the small contribution of mesoscale oscillations in the total variability of radiation when measuring the radiative properties of the earth as a planet, it is expedient to limit oneself to large-scale averaging over 1000 km or more.

BIBLIOGRAPHY

1. Boldyrev,V.G. et al. Korrelyatsionnye funktsii i spektral'nye plotnosti intensivnosti ukhodyashchego izlucheniya v intervale spektra 0.6 — 0.8 mkm (po izmereniyam so sputnika "Kosmos-45") (Correlation Functions and Spectral Densities of the Intensity of Emitted Radiation in the 0.6 — 0.8μ Range of the Spectrum (According to Cosmos 45 Measurements)). — Trudy Gidromettsentra SSSR, No. 36. 1968.
2. Kondrat'ev,K.Ya. et al. Pole izlucheniya Zemli kak planety (Radiation Field of the Earth as a Planet). Leningrad, Gidrometeoizdat. 1967.
3. Lebedinskii,A.I. et al. Korrelyatsionnye zavisimosti spektral'noi intensivnosti infrakrasnogo izlucheniya Zemli v Kosmos (Correlations of Spectral Intensities of Infrared Radiation of the Earth into Space). — Geomagnetizm i Aeronomiya, Vol. 8, No. 1. 1968.
4. Sonechkin,D.M. and I.S. Khandurova. Rezul'taty issledovaniya prostranstvenno-vremennoi izmenchivosti oblachnosti na Evropeiskoi territorii SSSR (Investigation Results of the Space-Time Variability of Cloudiness over European USSR). — Trudy Gidromettsentra SSSR, No. 50. 1969.

UDC 551.507.362.2:551.571.7

ASSESSMENT OF CHARACTERISTICS OF THE ATMOSPHERIC MOISTURE DISTRIBUTION WITH THE AID OF SATELLITE MEASUREMENTS

V. G Boldyrev, A. S. Gurvich, E. P. Dombkovskaya, T. G. Krasil'nikova

This paper investigates the possibility of using satellite radiometric measurements to recon-struct vertical moisture profiles. It is shown that satellite measurements of the total moisture con-tent yield data on the vertical distribution of specific humidity $q(z)$, if one makes use of a priori information in the form of statistical characteristics. Data are given on experimental verifications of the indicated methods, conducted on the basis of Cosmos 243 measurements.

Measurements of emitted heat radiation made by the Cosmos 243 satellite /1/ enabled the distribution and range of water vapor con-tent $Q_\Sigma = \int \rho(z)dz$ in the atmosphere to be determined over the ocean ($\rho(z)$ is the density of water vapor at height z). It has been shown that the general character of the water vapor distribution is con-nected with the motion of air masses /7/.

In this paper we demonstrate that satellite measurements of Q_Σ can be used to obtain data on the vertical distribution of specific humidity $q(z)$ at height z, if we use appropriate statistical properties: mean* profiles $\overline{q}(z)$ and the expansion $q' = q - \overline{q}$ with respect to statistical orthogonal functions /6/. Obukhov /6/ has shown that an optimum approximation of random functions, such as $q'(z)$, is effected by a system of orthogonal functions, being the eigenfunc-tions corresponding to the correlation functions $B_{qq} = q'(z_i) \cdot q'(z_j)$; given $q'(z)$ at a discrete number of levels this will be the basis of the system of vectors formed by the eigenvectors of the correlation matrices B_{qq},

$$q'(z_i) = q(z_i) - \overline{q}(z_i) = \sum_k a_k \varphi_{ik}, \tag{1}$$

where the expansion coefficients a_k can be regarded as random numbers for which $\overline{a_k} = 0, \overline{a_k^2} = \lambda_k$, λ_k being the eigenvalue of matrix

---

* The line above a quantity signifies the mean.

B_{qq} and φ_{ik} its eigenvector. For this reason the problem of determining the vertical moisture profile can be reduced to finding coefficients a_k from satellite measurements of emitted radiation /5/. It has been shown that all $q'(z)$ can be approximated with a high degree of accuracy by using expansions involving a few $(2-4)$ vectors /3/.

The limited accuracy of measurements of emitted radiation in radiowave lengths 8.5, 3.4 and 0.8 cm did not yield any information concerning the atmospheric water vapor content, and therefore these data were used to determine cloudiness and radiation from the underlying surface. Information on water vapor was obtained by observations on the 1.35 cm wavelength, a resonance length of water vapor.

Consequently, it was possible to find only one coefficient in expansion (1). We now show that the determination of even one coefficient can give important information about the profile of $q(z) = \overline{q}(z) + q'(z)$. We multiply (1) by $\Delta p_i/g$, where $p(z)$ is the pressure at height z, $\Delta p_i = p(z_{i+1}) - p(z_i)$, g is the acceleration of gravity, and sum over i. Hence

$$Q_\Sigma - \overline{Q}_\Sigma = \frac{1}{g} \sum_{k=1}^{n} a_k \Phi_k. \qquad (2)$$

In (2) Q_Σ is the mean value of the integrated moisture content $(\overline{Q}_\Sigma = -\frac{1}{g} \sum_i \overline{q_i} \Delta p_i)$, and $\Phi_k = -\sum_i \varphi_{ik} \Delta p_i$. The influence of the first term of (2) is now estimated. Since a_k is a random variable with mean value $\overline{a}_k = 0$, it is necessary to compare $\overline{a_1^2 \Phi_1^2}$ with $\overline{\left(\sum_{k=2}^{n} a_k \Phi_k \right)^2}$. Limiting ourselves to $n = 3$ and bearing in mind that $\overline{(a_2 \Phi_2 + a_3 \Phi_3)^2} \leqslant$ $\leqslant 2(\overline{a_2^2 \Phi_2^2} + \overline{a_3^2 \Phi_3^2})$, we see that relationship

$$\frac{\overline{(a_2 \Phi_2 + a_3 \Phi_3)^2}}{\overline{a_1^2 \Phi_1^2}} \leqslant \frac{2\left(\lambda_2 \Phi_2^2 + \lambda_3 \Phi_3^2 \right)}{\lambda_1 \Phi_1^2} \qquad (3)$$

gives an estimate of the first term of (2). If the characteristic values $\lambda_1, \lambda_2, \lambda_3$ are used in (3) (see Table 5), the variance of the first term on the right side of (2) is seen to be at least 10 times greater than the variances of the remaining terms. Therefore we can write the equality

$$Q_\Sigma - \overline{Q}_\Sigma = \frac{1}{g} \tilde{a}_1 \Phi_1. \qquad (4)$$

Determining Q_Σ from satellite measurements and knowing matrices B_{qq}, we can derive an estimate of the first coefficient of expansion (1).

The above estimates should not be regarded as conclusive, but the approach developed here allows the following to be stated: if the variance of the difference $a_1 - \tilde{a}_1$, equal to $\overline{(a_1 - \tilde{a}_1)^2}$, is considerably less than the variance of $\tilde{a}_1$, equal to $\overline{a_1^2} = \lambda$, then the estimate of $\tilde{a}_1$ obtained by solving equation (4) will give information about the profile of $q(z)$.

The effectiveness of the above method will depend on the selection of the expansion coefficient (and therefore of the autocorrelation matrix of moisture content). There arises the question of the variability of the autocorrelation matrices of moisture content (or rather of their eigenvalues and eigennumbers) as dependent on geographic and meteorological conditions, season, etc.

Many works /2, 4, 8/ mention the considerable variability of B_{qq} and other statistical properties of the moisture content, as dependent on the actual conditions for which they are computed, such as geographical latitude, state of atmospheric circulation, season of the year, etc. The conclusion reached is that the use of correlations, not related to actual conditions, for establishing vertical moisture profiles can lead to considerable errors.

It is known from /1/ that determination of the atmospheric moisture content from satellite radiometric measurements is at present possible only over water surfaces. Besides, considering the above-mentioned dependence of B_{qq} on actual physical and geographical conditions and the season, the statistical characteristics of specific humidity ($\overline{q}$, σ_q, B_{qq} and r_{qq}) were computed in our work using radiosonde observations from the North Atlantic weather ships A (62°N, 33°W), C (52°N, 35°W), D (44°N, 41°W) and E (35°N, 48°W) for different seasons (end of spring — beginning of summer; May, June; summer: July, August; fall: September, October) in clear weather (r_{qq} are the elements of the normalized matrices, the correlation coefficients).

Since clear weather over the North Atlantic (particularly north of 50°N) is rare, the number of vertical moisture profiles according to data from ships A and C was insufficient for statistical processing. However, since there are only insignificant differences in the physical and geographical conditions of atmospheric circulation between the regions of ships A and C, we combined their data into one statistical series.

To reveal the influence of latitude only on r_{qq}, the correlation matrices were computed separately for each region (regions of

TABLE 1

p_i, mb	1000	850	700	500	$rQ_{\Sigma q_i}$
Ships A and C, 56 cases					
1000	1.00	0.60	0.29	0.10	0.71
850		1.00	0.34	0.06	0.82
700			1.00	0.52	0.74
500				1.00	0.47
Ship D, 70 cases					
1000	1.00	0.66	0.33	0.14	0.75
850		1.00	0.40	0.12	0.80
700			1.00	0.42	0.69
500				1.00	0.41
Ship E, 94 cases					
1000	1.00	0.70	0.40	0.22	0.73
850		1.00	0.58	0.36	0.86
700			1.00	0.56	0.80
500				1.00	0.57
May — June, 54 cases					
1000	1.00	0.78	0.45	0.22	0.83
850		1.00	0.54	0.33	0.86
700			1.00	0.53	0.72
500				1.00	0.48
July — August, 96 cases					
1000	1.00	0.83	0.44	0.32	0.87
850		1.00	0.53	0.33	0.90
700			1.00	0.44	0.71
500				1.00	0.47
September — October, 42 cases					
1000	1.00	0.82	0.53	0.46	0.90
850		1.00	0.58	0.30	0.88
700			1.00	0.68	0.80
500				1.00	0.66

ships A and C, of ship D and of ship E) for all six months. To reveal the seasonal variation in r_{qq}, the correlation matrices were calculated using data of all four ships separately for each season.

Altogether 220 cases were selected for three years (1965 — 1967).

The analyses of normalized autocorrelation matrices (Table 1)
showed that the variability of elements of the correlation matrices
depended only to a small degree on geographical latitude over the
ocean.

TABLE 2

p_i, mb	1000	850	700	500	p_i, mb	1000	850	700	500
	Ship A					Ship D			
1000	1.45				1000	6.96			
850	0.61	0.73			850	3.59	3.50		
700	0.34	0.33	0.36		700	1.89	1.85	1.63	
500	0.09	0.07	0.06	0.04	500	0.51	0.41	0.32	0.16
	Ship C					Ship E			
1000	2.03				1000	10.32			
850	1.25	1.81			850	4.16	5.18		
700	0.75	0.93	0.97		700	1.77	1.94	2.37	
500	0.18	0.21	0.18	0.08	500	0.39	0.40	0.36	0.28

Absolute values of r_{qq} over the ocean differed little from those
over land.

TABLE 3

Ship A			Ship C			Ship D		
λ_1	λ_2	λ_3	λ_1	λ_2	λ_3	λ_1	λ_2	λ_3
1.94	0.44	0.17	3.71	0.75	0.37	10.1	1.6	0.47
φ_1	φ_2	φ_3	φ_1	φ_2	φ_3	φ_1	φ_2	φ_3
+0.82	—0.57	+0.02	+0.65	—0.75	—0.08	+0.80	—0.59	+0.10
+0.49	+0.69	—0.53	+0.63	+0.49	+0.59	+0.52	+0.59	—0.61
+0.28	+0.44	+0.83	+0.39	+0.43	—0.79	+0.29	+0.53	+0.76
+0.06	+0.06	+0.11	+0.08	+0.07	—0.11	+0.07	+0.07	+0.15

The nonnormalized correlation matrices differ a little more,
but not too much with regard to their structure. As an example
Table 2 shows nonnormalized correlation matrices for the above
four weather ships A, C, D, E for January. Two to three hundred
cases were used to calculate the matrices.

Table 3 gives values for the first three eigenvalues (λ_1, λ_2, λ_3) and eigenvectors (φ_1, φ_2, φ_3) of the matrices of Table 2 for Ships A, C, D, and shows that the variability of even the first eigenvector is fairly noticeable.   Application of the above method for determining the total moisture content must give better results in regions and seasons with high average tropospheric moisture and with large variability of moisture content.

TABLE 4

p_i, mb	φ_1	φ_2	φ_3	p_i, mb			
				1000	850	700	500
			Ship D				
1000	0.79	—0.60	0.10	7.35			
850	0.57	0.68	—0.44	3.93	4.66		
700	0.20	0.41	0.87	1.27	1.22	1.75	
500	0.03	0.05	0.17	0.21	0.15	0.27	0.22
			Ship E				
1000	0.65	—0.71	0.28	5.34			
850	0.67	0.35	—0.65	3.49	5.26		
700	0.35	0.60	0.70	1.56	2.22	2.63	
500	0.07	0.12	0.10	0.27	0.46	0.47	0.25

For an experimental check of the method of determining the moisture content we took 30 measurements of Q_Σ obtained on the Cosmos 243 satellite, for which radiosonde measurements also existed.   These data were divided into two groups, for latitudes north and south of 44°N.   The radiosonde profiles were used to calculate values of coefficients a_i.   Computations for the first group involved mean profiles and matrix B_{qq} calculated for Ship D (44°N) for May — October; for the second group corresponding data for Ship E (35°N) was used for the whole period, as listed in Table 4.   This table also tabulates the three first eigenvectors of the matrices.   The eigennumbers and eigenvalues are given in Table 5.

TABLE 5

Weather ships	λ_1	λ_2	λ_3	Φ_1	Φ_2	Φ_3	$\dfrac{2\left(\lambda_2\Phi_2^2 + \lambda_3\Phi_3^2\right)}{\lambda_1\Phi_1^2}$
D	10.52	1.97	1.32	250	104	160	0.10
E	9.82	2.21	1.27	220	90	104	0.06

TABLE 6

	Station No.																$\sum (a_1 - \tilde{a}_1)^2$
	1	2	3	4	5	6	7	8	9	10	11	12	13	14	15	16	
First Group																	
Latitude, deg	66	55	59	52	66	53	56	55	59	71	62	46	62	55	46	71	
a_1	−6.80	−4.98	−0.64	−3.13	−5.17	−3.43	−1.04	−2.82	−0.94	−5.70	−5.60	−4.27	−4.62	−4.17	−2.72	−6.86	0.25
$\tilde{a}_1$	−6.10	−3.95	−1.16	−2.75	−4.75	−2.90	−0.89	−2.37	−0.61	−5.26	−5.80	−4.11	−5.26	−3.80	−2.37	−7.39	
Second Group																	
Latitude, deg	17	13	7	−37	18	9	28	−21	−37	−29	18	18	13	18			
a_1	8.26	8.39	9.82	0.73	6.13	7.14	6.77	2.14	−3.51	−1.57	6.13	6.76	8.37	4.05			3.59
$\tilde{a}_1$	10.62	9.28	11.58	1.32	5.87	8.38	8.38	5.62	0.28	0.32	6.10	5.31	7.91	2.05			

This layout and choice of time allow us to assume that our data belong to a set possessing statistical properties described by the mean profiles and matrices B_{qq} used by us.

Table 6 gives values of coefficients a_1 and $\tilde{a}_1$ obtained from radiosonde observations. The values of a_1 were found by expanding the profiles $q(p_i)$ with respect to vectors φ_{ik} ($k = 1, 2, 3$) using Table 4. The values of $\tilde{a}_1$ were determined by solving equation (4) for known $\overline{Q_\Sigma}$ and Q_Σ measured from a satellite. The estimate of the variance $\overline{(a_1 - a_1)^2}$ for all groups is given in <u>Table 6</u>. A comparison indicates that the experimental values of $(a_1 - \tilde{a}_1)^2$ are at least three times less than the corresponding values of λ_1 given in Table 5. Consequently these experimental values of $\tilde{a}_1$ yield information on the vertical moisture profile. As expected, $\overline{(a_1 - \tilde{a}_1)^2} \approx \lambda_2 + \lambda_3$, i. e., the error in determining $\tilde{a}_1$ by the above method is of the order of the discarded terms (starting from $k = 2$) of the sum (1).

The above results indicate that satellite measurements of the integrated moisture Q_Σ yield an additional representation of the vertical distribution of moisture in the atmosphere over the ocean, at least in specified geographical regions and seasons.

BIBLIOGRAPHY

1. Basharinov, A. E., A. S. Gurvich, and S. T. Egorov. Opredelenie geofizicheskikh
      parametrov po izmereniyam teplovogo radioizlucheniya na ISZ "Kosmos-243" (De-
      termination of Geophysical Parameters on the Basis of Cosmos 243 Heat Radiation
      Measurements). — DAN SSSR, Vol. 188, No. 6. 1969.
2. Boldyrev, V. G. and V. M. Oleshev. O statisticheskoi strukture vertikal'nykh profilei
      temperatury i vlazhnosti (Statistical Structure of Vertical Temperature and Moisture
      Profiles). — Trudy MMTs, No. 11. 1966.
3. Boldyrev, V. G., L. I. Koprova, and M. S. Malkevich. Ob uchete variatsii vertikal'nykh
      profilei temperatury i vlazhnosti pri opredelenii temperatury podstilayushchei pover-
      khnosti po ukhodyashchemu izlucheniyu (Computation of Variations in Vertical Tempera-
      ture and Moisture Profiles for the Determination of the Temperature of the Underlying
      Surface with the Aid of the Emitted Radiation). — Izvestiya AN SSSR, Seriya Fizika
      Atmosfery i Okeana, Vol. 1, No. 7. 1965.
4. Komarov, V. S. Nekotorye statisticheskie kharakteristiki vertikal'nykh profilei temperatury
      i vlazhnosti (Some Statistical Properties of Vertical Temperature and Moisture Profiles). —
      Trudy NIIAK, No. 58. 1969.
5. Malkevich, M. S. and V. I. Tatarskii. Opredelenie profilya temperatury atmosfery
      po ukhodyashchemu izlucheniyu (Determination of Atmospheric Temperature Profiles
      from the Emitted Radiation). — Kosmicheskie Issledovaniya, Vol. 3, No. 3. 1965.
6. Obukhov, A. M. O statisticheskikh ortogonal'nykh razlozheniyakh empiricheskikh funktsii
      (Statistical Orthogonal Expansions of Empirical Functions). — Izvestiya AN SSSR, Seriya
      Geofizicheskaya, No. 3. 1960.

7. Obukhov,A.M. and M.S. Tatarskaya. Pole integral'nogo vlagosoderzhaniya atmosfery nad yuzhnym polushariem po izmereniyam teplovogo radioizlucheniya na sputnike "Kosmos-243" (The Integrated Atmospheric Moisture Field over the Southern Hemisphere Determined from Cosmos 243 Heat Radiation Measurements). — Meteorologiya i Gidrologiya, No.11. 1969.
8. Popov,S.M. Nekotorye statisticheskie kharakteristiki vertikal'noi struktury polei temperatury i vlazhnosti (Some Statistical Properties of the Vertical Structure of Temperature and Moisture Fields). — Izvestiya AN SSSR, Seriya Fizika Atmosfery i Okeana, Vol.1, No.1. 1965.

UDC 551.507.362.2:551.571.7

INTERRELATIONSHIP BETWEEN DEVELOPMENTS OF SYNOPTIC PROCESSES AND EVOLUTION OF THE INTEGRAL MOISTURE FIELD ACCORDING TO SATELLITE MEASUREMENTS

E. P. Dombkovskaya, V. V. Demin

Results of simultaneous analyses of synoptic maps and maps of integral moisture are presented. It is shown that the development of synoptic processes is reflected in a specific way by the evolution of the integral moisture field and by the vertical redistribution of water vapor.

The possibility of a synoptic representation of satellite data on the total atmospheric moisture content (integral moisture) $w^* = \int_0^\infty \rho_w\, dz$ has been discussed in /2/, and an example of such an interpretation was given in /1/. These synoptic interpretations had a statistical character: as a result of this interpretation a connection was established between different synoptic phenomena and elements of the field of w^*. The dynamics of synoptic processes and their reflection in the field of w^* was not treated in these papers. However, since in the analysis and prognosis of synoptic situations the dynamics of processes are of paramount importance, we consider it expedient to study the interrelationship of synoptic processes and the resulting changes in the integral moisture field.

In this article we present results of simultaneous analyses of synoptic situations and of the moisture field, constructed in accordance with radiometric measurements from Cosmos 243 over the Atlantic Ocean for three consecutive days (23 − 25 September 1968). In addition to these measurements and synoptic maps, we used upper-level maps and nephanalyses as well as British daily weather reports /4/ and radiosonde observations from 9 weather ships in the North Atlantic. The integral moisture maps were constructed as in /1/.

Synoptic processes over the North Atlantic, observed at the beginning of the examined period, can be attributed to two circulation types after Sorkina /3/. They characterize the initial stages of breakup of a subtropical anticyclone from the south, owing to

activation of the trade wind fronts.   The actual synoptic process in
the examined three days developed as follows.

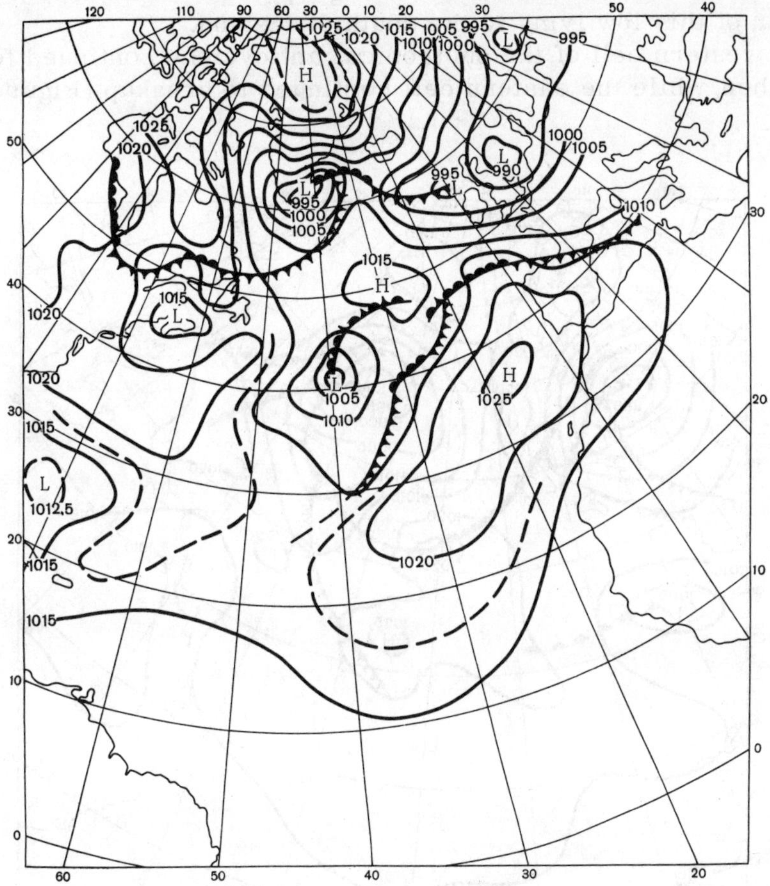

FIGURE 1.   Synoptic map at 1500 hrs Moscow time on 23 September 1968.

A depression with two centers, occupying the central Atlantic,
was filling.   At the same time one of its centers, lying off the
southern end of Greenland, moved slowly east while the second
center in northern England remained quasistationary.   An occluding
depression from the Azores moved northeast while filling.   A depression from the Greater Antilles did not change in intensity and
moved slowly north, breaking down the southern periphery of the
subtropical anticyclone.   The western cell of the subtropical anticyclone moved east while strengthening and expanding in area;
at the same time its eastern cell, oriented northwest to southeast,
remained quasistationary and disintegrated (Figure 1).

At 1500 hrs Moscow time on 25 September the depression north-east from the Azores merged with the northern depression, and this led to the regeneration and formation of a deep low with pressure at the center below 995 mb.   A small high-pressure cell appeared at the rear of this low lying to the south of Iceland.

The western cell of the subtropical anticyclone continued to strengthen, while the eastern cell continued to breakup (Figure 2).

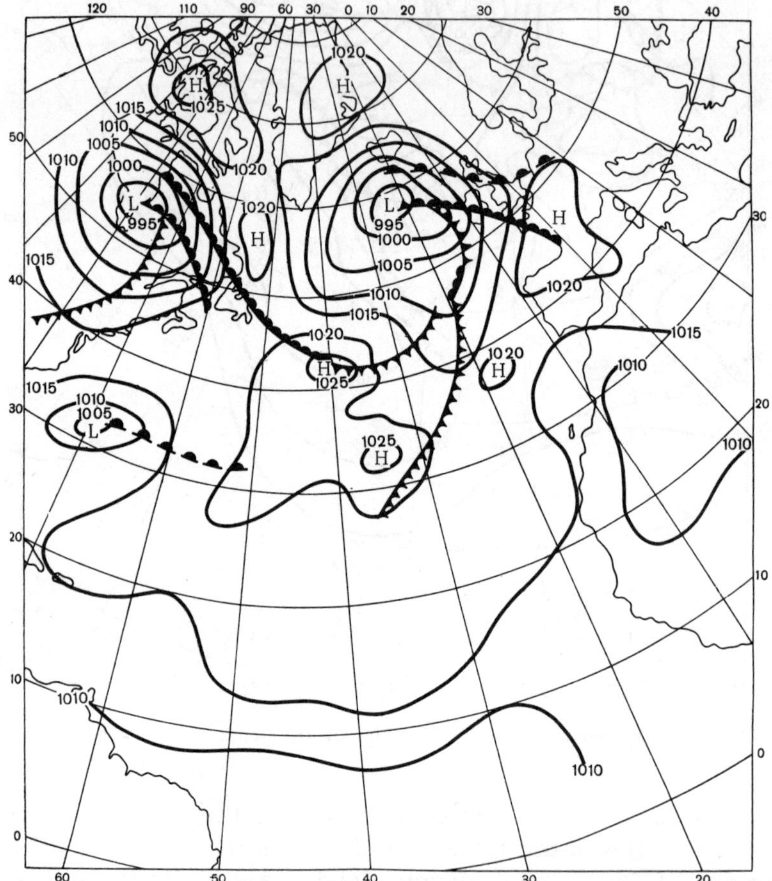

FIGURE 2.   Synoptic map at 1500 hrs Moscow time on 25 September.

The evolution of the field of integral moisture over the North Atlantic accompanying the indicated development of synoptic processes was as follows (Figures 3 and 4).   A center of relatively dry air over the western cell of the subtropical anticyclone (marked on the maps by the letter A) expanded together with the latter and

moved slowly east.   A second, relatively dry pocket (B) corresponded to the eastern cell of the subtropical anticyclone, decreased
in area, and remained almost stationary like the anticyclone.   A
narrow tongue of moist air (C), connected with a frontal surface,
extended from the British Isles to the southwest and moved together with the front.   On nephanalysis maps a well marked frontal
cloud band corresponded to this moist tongue.   The development of
the centers of maximum moisture (5 to 5.5 g/cm³), marked D, in
the subequatorial regions (0 − 10°N) reflected the evolution of
cloudiness in the tropical convergence zone.

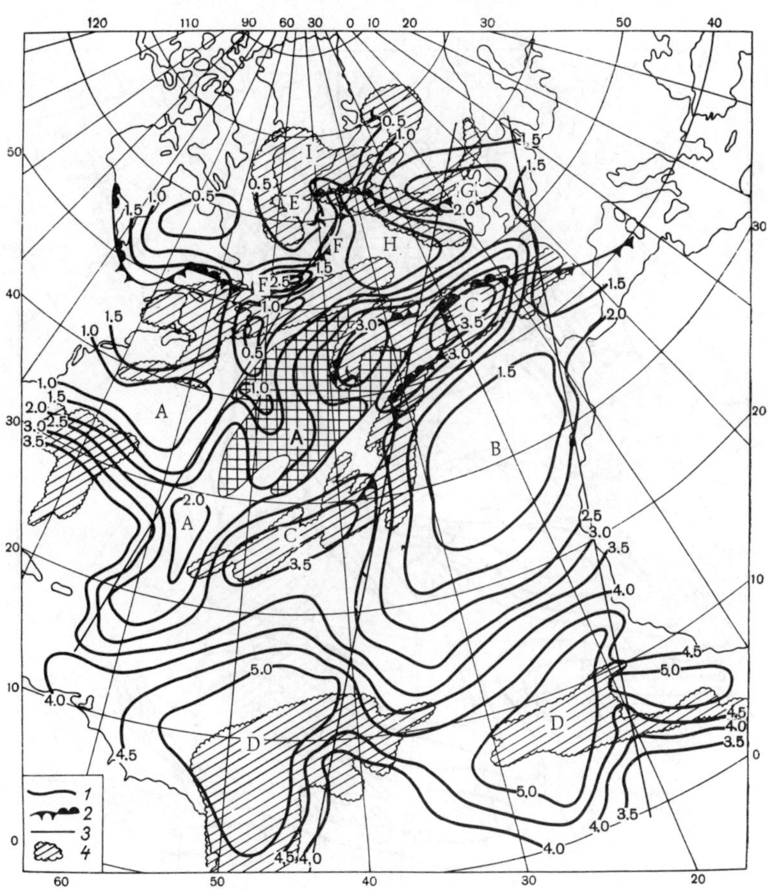

FIGURE 3.   Map of integral moisture over the North Atlantic Ocean on 23
September about 1500 hrs Moscow time:

1 — isolines; 2 — fronts; 3 — projection of satellite passes; 4 — limits of cloud
fields (nephanalysis data).

A more complex evolution was seen in the integral moisture field in the northern part of the examined ocean area (north of 50°N). On the initial day, 23 September, 1500 hrs Moscow time, the field of w^*, connected with the two-centered depression, presented a very variegated picture. On the moisture chart a center (E) of dry air ($w^* = 0.5$ g/cm^3) corresponded to the depression lying at the south coast of Greenland. To the south and east this dry pocket was surrounded by narrow belts of moist air (F), coinciding with a frontal surface. Over the second center of the low over northern England lay a moist air pocket (G) ($w^* = 2.0$ g/cm^3), and at its rear, in the high-pressure cell, lay a small dry core (H), oriented meridionally and extending from the large center of dry air (I) over Greenland.

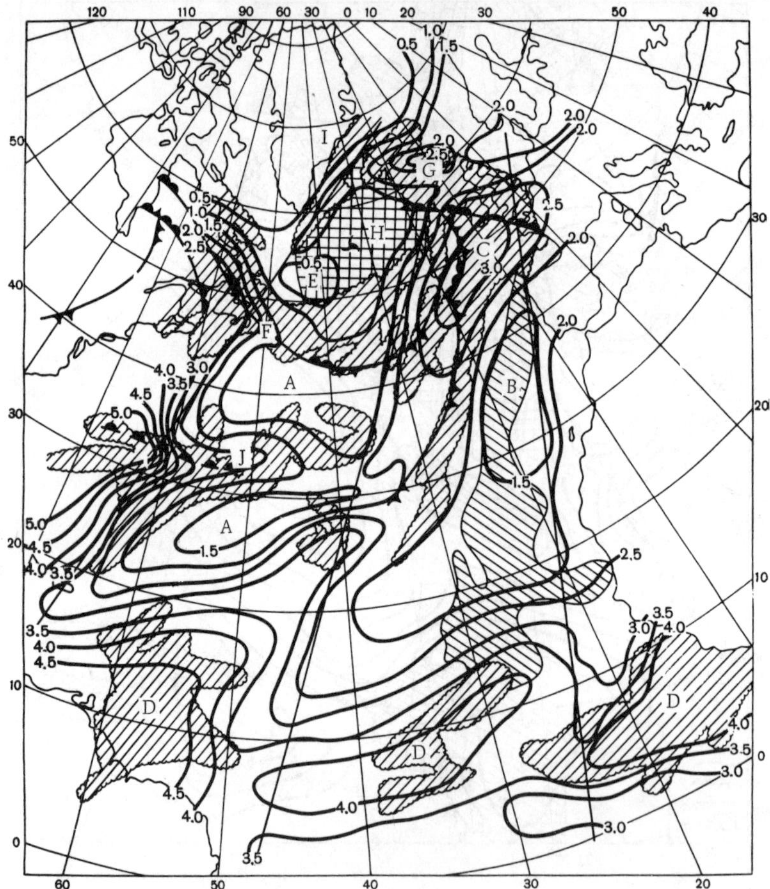

FIGURE 4. Map of integral moisture over the North Atlantic Ocean on 25 September about 1500 hrs Moscow time:

1 — isolines; 2 — fronts; 3 — projection of satellite passes; 4 — limits of cloud fields (nephanalysis data).

With the further development of the synoptic process of filling of
the depression center and the breakup of its connected front, as well
as the merging of the southern low with the eastern center of the
depression and its subsequent deepening and regeneration, the moisture field also changed.   The dry air pocket (E) widened, displacing
southward the moist belt (F) which then broke up.   On 25 September,
1500 hrs Moscow time, the dry cell (E) occupied the entire central
North Atlantic.   To the east and southeast of this vast dry center
lay regions of moist air (C and G), coinciding with the frontal part
of the depression and an upper warm front, extending from Denmark
through northern England to Iceland.

A comparative analysis of aerological data and integral moisture
maps confirmed the ideas expressed in /1/ regarding the dependence
of the total atmospheric moisture content on the nature of vertical
motions in the air, and the different synoptic formations with certain
elements of the integral moisture field w^*.   It also enables some
conclusions to be drawn concerning the relationship between the
movement of synoptic formations and their dynamic transformations,
and the evolution of the integral moisture field.

In this connection the following points are noteworthy.

1) Anticyclones and ridges, which are characterized by descending air motion, are associated with centers of low moisture; this
holds true since low moisture is typical not only for anticyclones
with clear skies or little cloudiness, but also with widespread
cloudiness beneath inversions: e. g., the center of dry air B in
Figure 3 lies in a zone of few clouds, whereas in Figure 4 it lies
in a zone of continuous stratiform clouds.

2) Frontal surfaces, where rising air motions are usually observed, correspond to narrow centers of high moisture.

3) The intertropical convergence zone with formations of Cb
clouds corresponds to centers of maximum moisture.

4) Depressions may correspond to centers of dry as well as of
moist air.   This is apparently explained by the existence in depressions of areas of both rising and descending motion (e. g., at
the rear of lows, and in filling depressions and in their central
parts).

5) Elements of the moisture field move together with their
associated synoptic formations.

Changes in synoptic systems (deepening or filling of lows,
strengthening or dissolution of anticyclones, changes in intensity,
etc.) are not obviously reflected in the field of integral moisture.
Thus, for instance, the strengthening (dissolution) of anticyclones
is not immediately accompanied by a marked decrease (increase)
in integral moisture: the eastern cell of the subtropical anticyclone

decayed noticeably during the three days of investigation, while its
central pressure decreased from 1025.3 to 1020 mb, and its area,
surrounded by a closed isobar, decreased markedly. At the same
time the pocket of dry air (according to data at our disposal and
the map of w^*) did not noticeably weaken. This is confirmed also
by values of the correlation coefficients r_{pw^*} (between surface
pressure p and total atmospheric moisture content w^*, calculated
for different weather conditions) which did not exceed 0.44.

Transformations occurring in frontal zones are somewhat more
clearly reflected on maps of integral moisture: simultaneously
with the disintegration of a front the associated narrow moist
tongue also dissolves (e. g., the strip of moist air F); conversely,
with an intensification or the appearance of a front there is a
strengthening or new appearance of a belt of moist air (e. g., strip
J on Figure 4, connected with the front of the depression near the
Antilles).

A specific property of the meteorological parameter w^* and its
dependence on the nature and intensity of vertical motions suggests
the possibility of a relationship between the vertical distribution
of water vapor (layer content) and the development of synoptic
processes.

During anticyclogenesis the accompanying intensification of
downward motion in the lower and upper levels of the troposphere
must increase in favor of the lower layers; conversely, with cyclogenesis and frontogenesis, the accompanying strengthening of upward motion of air masses must increase in the upper layers.
In practice this expresses itself in changes of values of the ratio
$\dfrac{\Delta w_i}{w^*} \cdot 100$, where Δw_i is the water vapor content in the i-th layer of
the troposphere, $i = 1$ is the $1000 - 850$-mb layer, $i = 2$ the
$850 - 700$-mb layer, and $i = 3$ the $700 - 500$-mb layer.

The analysis of aerological data given here confirmed the correctness of this view. As an example we examine the change in
the vertical distribution of water vapor in the regions of ships D
and J, where the synoptic transformations are particularly pronounced. Table 1 lists data for the layer content of water vapor
in these regions on 23, 24 and 25 September, 1500 hrs Moscow
time, calculated from radiosonde measurements.

These data show that in the region of ship D, in the lowest
1.5 km layer, the relative humidity increased over the three days
from day to day (60.5, 65.7 and 70.5%), while in the upper layers
it decreased (e. g., in the $700 - 500$-mb layer from 12.0% on
23 September to 7.6% on 25 September). During these three days
the synoptic situation in the region of ship D changed as follows:

on 23 September the region of ship D lay on the periphery of a
depression, on 24 September it lay on the periphery of a ridge and
on 25 September in an anticyclone.

TABLE 1. Layer content of water vapor in the regions of ships D and J

	Ship D			Ship J		
	23 Sept.	24 Sept.	25 Sept.	23 Sept.	24 Sept.	25 Sept.
w^* g/cm^2	1.66	1.54	1.32	1.14	1.73	1.34
Δw_1 g/cm^2	1.01	1.01	0.93	0.70	0.83	0.82
$\dfrac{\Delta w_1}{w^*} \cdot 100\%$	60.5	65.7	70.5	61.5	48.0	61.1
Δw_2 g/cm^2	0.40	0.36	0.25	0.30	0.48	0.37
$\dfrac{\Delta w_2}{w^*} \cdot 100\%$	24.0	23.4	18.9	26.3	27.7	28.6
Δw_3 g/cm^2	0.20	0.13	0.10	0.09	0.29	0.12
$\dfrac{\Delta w_3}{w^*} \cdot 100\%$	12.0	8.5	7.6	7.9	16.7	9.0

On 23 and 24 September the forward part of a depression approached the region of ship J, and consequently the relative humidity
in the surface layer decreased from 61.5 to 48.0%, while in the
700 — 500-mb layer it increased from 7.9 to 16.7%. On subsequent
days, when the region of this ship lay at the rear of a depression
and behind the cold front, the relative humidity rose in the surface
layer to 61.1%, whereas in the upper layers it dropped to 9.0%.

The above property of w^* and Δw_i has, in our opinion, great possibilities for weather analysis and prognosis for data-sparse
ocean areas of the earth.

In conclusion, we wish to point out that since the amount of
data from Cosmos 243 was very limited, the results of the analysis
presented here are regarded by us as preliminary, and this article
is only an example of the synoptic interpretation of radiometric
satellite measurements.

BIBLIOGRAPHY

1. Gurvich, A.S., V.V. Demin, and E.P. Dombkovskaya. Ispol'zovanie sputnikovykh
      kart obshchego vlagosoderzhaniya v sinopticheskom analize (Use of Satellite Maps of the
      Total Moisture Content in Synoptic Analysis). — Meteorologiya i Gidrologiya, No. 8. 1970.

2. O b u k h o v , A. M. and M. S. T a t a r s k a y a . Pole integral'nogo vlagosoderzhaniya atmosfery
    nad yuzhnym polushariem po izmereniyam teplovogo radioizlucheniya na sputnike
    "Kosmos-243" (Integral Atmospheric Moisture Content Field over the Southern Hemisphere
    According to Cosmos 243 Heat Radiation Measurements).— Meteorologiya i Gidrologiya,
    No. 11. 1969.
3. S o r k i n a , A. I. Tipy atmosfernoi tsirkulyatsii i vetrovykh polei nad severnoi chast'yu
    Atlanticheskogo okeana (Types of Atmospheric Circulation and Wind Fields over the
    North Atlantic Ocean). — Trudy GOIN, No. 84. 1965.
4. Daily Weather Report of the British Meteorological Office, Vol. 9. 1968.

UDC 551.509

COMPARISON OF SATELLITE-MEASURED
AND RADAR-OBSERVED CLOUD-TOP HEIGHTS

G. I. Il'ina, V. F. Lapcheva

Calculated cloud-top heights are compared with those obtained from radar observations for the summer of 1970. On the average, the values calculated from satellite data are found to be 1 km less than those determined from radar measurements. A brief presentation is given of the technique of utilizing maps of cloud-top heights in nephanalysis.

In February 1970, the USSR Hydrometeorological Center began to regularly issue cloud-top heights, which were used in nephanalysis practice. The heights are computed by a program /1/ utilizing radiation temperatures, measured by actinometric apparatus in the $8-12\,\mu$ window region.

Cloud-top heights provided by the Meteor satellite were compared with radar and radiosonde data for the summer of 1970. Satellite and radar data on cloud fields (Figure 1) were viewed simultaneously at the USSR Hydrometeorological Center, the area viewed by the meteorological radar ($R = 300\,$km from Moscow) coinciding with the central part of the satellite coverage (50–60°N, 32–42°E). The duration of the data-gathering period was 5 min for the satellite and approximately 15–20 min for the radar (with the screen completely filled). The maximum deviation in the duration of radar and satellite observations in the cases at hand were 28 min, with a mean of 15 min.

It was not always possible to carry out observations with a higher degree of synchronization, since the radar of the USSR Hydrometeorological Center is primarily employed in providing radar data to the laboratory of short-term weather forecasts.

In comparing satellite and radar actinometric information, consideration must be given to the specific measurement techniques. Radar information is issued in $30 \times 30\,$km squares, while actinometers provide for $50 \times 50\,$km resolution at the nadir. Increasing the scanning angle results in poorer resolution. However, since data analyzed in this paper pertain to the central part of the

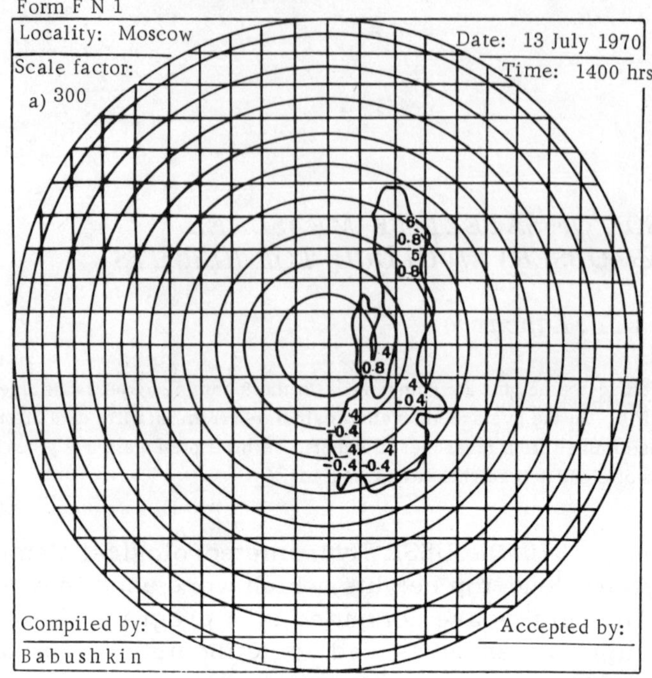

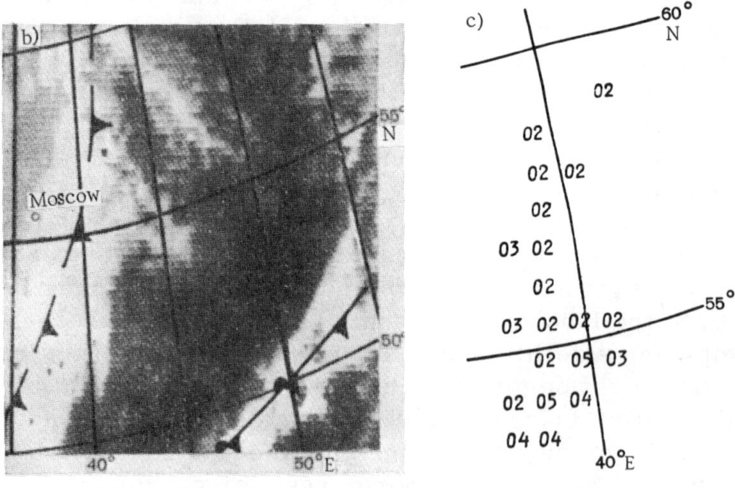

FIGURE 1:

a — radar photograph; b — infrared photographs; c — cloud-top height map.

scanning band, it can be assumed that the resolving power of the apparatus over this section does not differ significantly from the above. The permissible accuracy in referencing satellite information is approximately 100 km.

The above approach allows single cloud-top heights obtained by satellite and radar data to be compared precisely. The comparison is hence carried out over the entire cloud mass, the outline of which is obtained from TV or infrared photographs. The cloud mass was subdivided into northern and southern regions (relative to Moscow), while the radar scanning area was subdivided into 150 X X 150 km squares. Predominating and maximum cloud-top heights were selected for these regions from cloud-top height maps and from their corresponding squares on the radar scan.

From June to September of 1970 we carried out 17 simultaneous satellite and radar observations. Analysis was based only on data for which qualitative observations were available concerning all kinds of satellite information. If information pertaining to only one aspect was absent, the data for that case were discarded. Radiosonde data for periods close to the time of satellite passage were also used (permissible time difference not exceeding 1 hour), as well as TV and infrared information obtained by direct transmission.

The results are listed in Table 1 for a total of 12 cases, of which the heights in eight cases obtained from satellite data were equal to, or 1–2 km lower than those determined from radar observations. On the average, the eight cloud-top heights obtained from the program are underestimated by 1 km.

If one considers the possible error in determining the radiation temperature, and also the fact that the difference between the cloud-top height obtained from satellite and radiosonde observations is about 1 km /1/, then agreement between radar and actinometer observations can be regarded as satisfactory.

Table 1 shows that cloud-top heights obtained from the radar and satellite data are those of medium-level clouds and clouds with vertical structure. It should be noted that the radiation of medium-level and high clouds is not that of an ideal blackbody. According to Novosel'tsev /2/, the "blackness" of medium-level clouds in the 8–12-μ window region varies from 0.40 to 0.73 along the normal, while that of high clouds varies from 0.16 to 0.26.

Emittance characteristics are disregarded in the program for calculating cloud-top heights. This can also serve as an important source of errors, particularly from the standpoint of detecting and determining the height of cirrus tops.

Aircraft observations of the radiation and actual temperature over and in jet-stream cirrus were carried out in the eastern half

TABLE 1. Comparison of cloud-top heights obtained by the present program with radar and radiosonde heights for the summer of 1970

Date	Time (hrs)		Orbit number	Cloud-top height H (km)			Difference (km)		Cloud form		Remarks
	satellite (sat)	radar (rad)		satellite	radar	radio-sonde (son)	$H_{sat}-H_{rad}$	$H_{sat}-H_{son}$	satellite	ground stations	
13 July	1400	1410	261	2–3	5	4	-2	- 1		Cb, Ac, Ci	Cloudiness of secondary cold front
13 July	1400	1410	281	4–5	4	10	1	- 5		Ac, As, Ci	The same
7 Aug.	2020	1955	637	3–7	8	10	-1	- 3		Ac, Cu, Cb	Cloudiness of slow-moving front
7 Aug.	2020	1955	637	3–10	12	—	0	—		Ac, As, Cu	ESSA-8 data as of 1139 hrs
29 Aug.	0759	740	940	2–4	3–4	8	0	- 4		Ac, As, Ci	Warm-front cloudiness
29 Aug.	0759	740	940	2–3	3–4	10	-1	- 7		Ac, As, Ci, Sc	The same
24 Sept.	1558	1553	1311	2–3	4	2	-1	1		Cb, Ci, Cu, Ac	Cloudiness of secondary cold front
24 Sept.	1558	1553	1311	2–5	3	9	2	- 4		Cb, Ci, Cu, Ac	The same
24 Sept.	1558	1553	1311	2–3	3–4	8	-1	- 5		Cb, Ci, Cu, Sc	" "
26 Sept.	1533	1505	1338	2–4	4	6	0	- 2		Cb, Ac	" "
26 Sept.	1533	1505	1338	3–5	4	—	1	—		Cb, Cu, Sc, Ac	" "
28 Sept.	1517	1528	1367	3–4	3	8	1	- 4		Cb, Cu, Ac, Ci	" "
Σ							-1	-34			
ΔH_{av}							-0.1	- 3			
Σ_{8}cases							-6				
ΔH_{av}							-1				

of the United States from April 1964 to February 1966 /4/. It was found that the average deviation of the actual from the radiation temperature at the level of cirrus cloud tops was 36°, and the mean lapse rate was 2.5°/300 m. Hence the height of cirrus cloud tops was 4.5 km underestimated.

The detection and localization of cirrus in space is in general a difficult problem. Radiosonde data, in particular, yield little additional information for the reliable interpretation of cirrus. Hence radiosonde data taken during satellite and radar observations were utilized for the approximate detection of high clouds from the dew-point deficit. The cloud-top height for cirrus determined from radiosonde data was not confirmed by satellite actinometric data or by radar on 13 July, 29 August, and 24 and 28 September, although the existence of cirrus was confirmed in every case by infrared photography.

Consider now the accuracy of radar measurements. The probability of detecting all cloud forms decreases with distance and depends on the rating of the radar set, the microstructure of the clouds and their vertical thickness. For example, the probability /3/ of detecting low clouds at a distance of 100—150 km is 45—35%, while for high clouds it is 5%. Vertical and cumulonimbus clouds are detected from a distance of 200 km with a probability of 95—100%, while the same clouds at a distance of 250—300 km are determined with a probability of 5—15%. The mean probable error /3/ in determining cloud-top heights of medium-level and low clouds at a distance of 30 km from the observation point is ± 150 m, while at a distance of 50 km the cloud-top height cannot be detected by radar. The height of cumulonimbus located at 150 km from the point of observation is determined to within ± 0.5 km, while at 200—300 km the accuracy is ± 1 km.

The insufficient probability of detecting some cloud forms can also serve as a source of differences between satellite and radar observations.

The technique of utilizing the cloud-top height in nephanalysis practice is now treated. Data on cloud-top heights can sometimes also be used as additional cloud-cover characteristics for refining the definition of cloud form in interpreting observational data.

The USSR Hydrometeorological Center utilizes these data in the following manner. The cloud-top heights are taken from the map only for major cloud systems (fronts, cloud spirals). The selection is made inside the cloud mass and involves finding the predominating and then the maximum heights. A single maximum is used when it does not differ from preceding values by more than 3 km. The heights thus determined are recorded in free space inside the cloud field of the nephanalysis map.   The data is recorded by a two-digit number, according to international requirements, and also in the form of a fraction, the numerator of which is the maximum height and the denominator the prevailing height.   For example, if the indicated height is 09/06 km, this means that the maximum cloud-top height is 9 km and the prevailing cloud-top height is 6 km. Heights of 10 km and more are represented by the symbol >.   It is standard practice to designate heights of 10 km and more on nephanalysis maps by the numeral 10.

Maps of cloud-top heights can also be used for other cloud masses;   however, the existence of high clouds requires a critical approach to these maps.   For example, if the cloud-top height is 4—5 km while infrared photography of this region confirms the presence of cirrus only, then the height values are not used.   As an example of applying cloud-top height maps to the interpretion of infrared photographs, we now analyze information in the Pacific Ocean region (5°N—5°S, 95—85°W) during orbit 986—987 on 1 September 1970 (Figure 2).

Infrared photograph, Figure 2b, illustrates a quite dense cloud mass located in this region.   The person interpreting the photograph could claim the presence here of cumulonimbus, cirrus and stratocumulus.   TV photograph, Figure 2a, confirms the presence only of stratocumulus, the cloud-top heights of which are 2—3 km, with a correction of 3—4 km.   According to these heights, there is no cirrus in this cloud mass and the clouds are primarily stratocumulus;   this conclusion is confirmed by the ground-analysis map. A weather ship recorded stratocumulus, altocumulus and altostratus.

Consider another example: analysis of information in the Pacific Ocean region (15—20°N, 110—115°E) during orbit 454—455 on 25 July 1970 (Figure 3).   The infrared photograph allows one to start immediately with cloud-form interpretation.   Some analysts may claim cumulus and cirrus with some cumulonimbus, others may find the clouds to be cumulus and stratocumulus, and still others only cumulus.   The TV picture in the region under study clearly shows a cumulus vortex pattern;   the cloud-top heights lie below 2 km.   In this case it is obvious that the actual cloud form is that of cumulus. Ground-observation maps confirm this by showing stratocumulus.

To the west of the region under study (115—120°W) the infrared and TV pictures show cumulonimbus, cirrus and cumulus, the upper boundary of which is as high as 9 km.   The prevalence of these cloud forms is confirmed by weather ships.   One ship recorded 10-point continuous stratus with heavy rains, while the other found cumulus, altocumulus and cirrus.

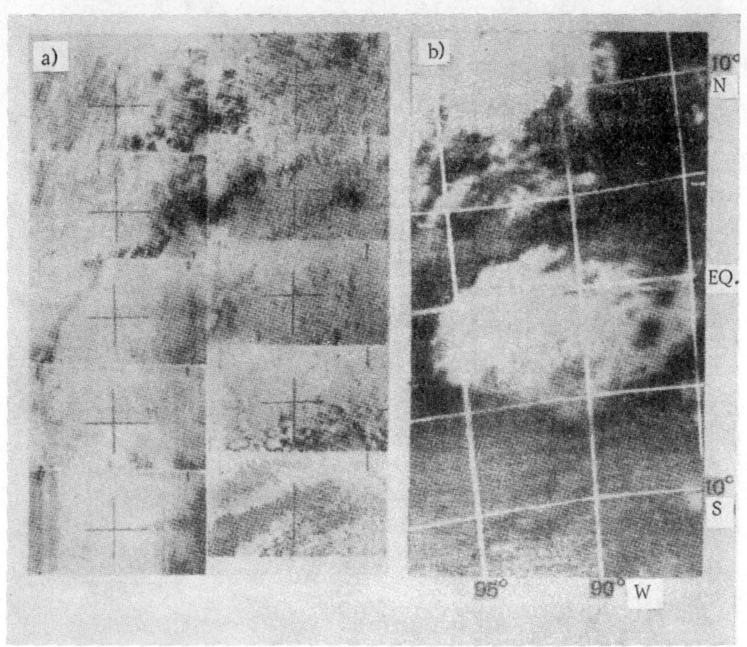

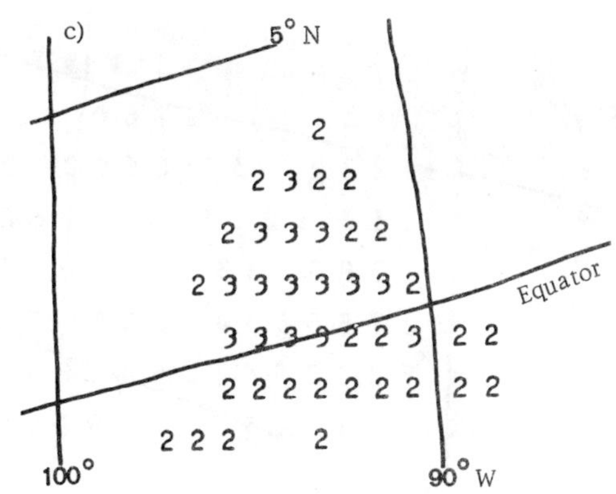

FIGURE 2:

a — TV photographs; b — infrared photographs; c — cloud-top height map.

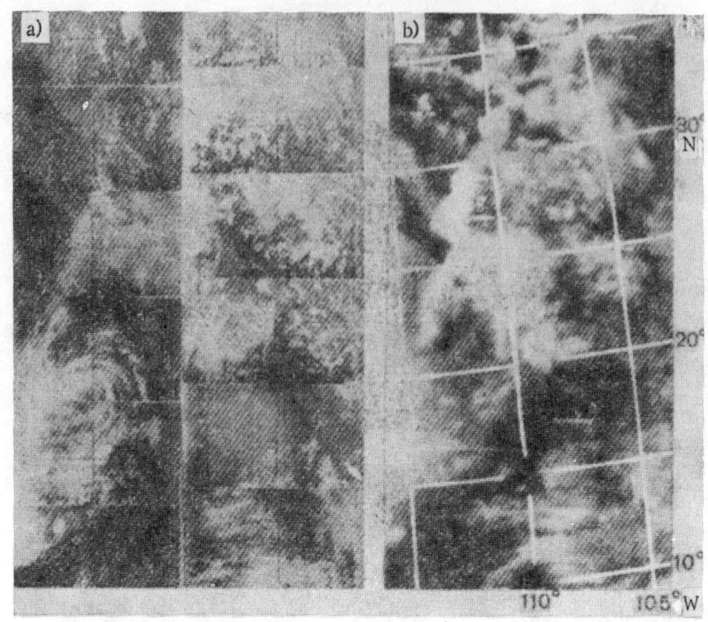

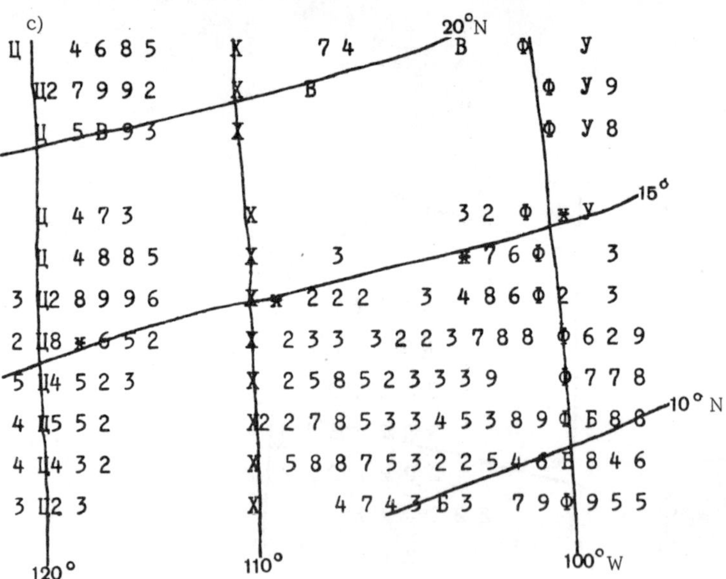

FIGURE 3:

a — TV photographs; b — infrared photographs; c — cloud-top
height map.

It would be possible to present other examples in which the cloud form is determined more precisely from infrared photography with the aid of cloud-top height maps.

The following conclusions may consequently be stated.

1. Cloud-top heights of medium-level and vertical clouds calculated from a program utilizing satellite-measured radiation temperatures are on the average 1 km lower than heights obtained from radar observations.

2. The radar is incapable of determining cloud-top heights of cirrus, while their values calculated from satellite-measured radiation temperatures are on the average 5 km underestimated. Hence in the presence of high clouds the cloud-top heights can be used only for a combination of dense cirrus with thick vertical clouds or with dense medium-level clouds.

BIBLIOGRAPHY

1. Boldyrev, V.G. and A.N.Katal'nikova. Nekotorye rezul'taty opredeleniya vysoty verkhnei granitsy oblakov po sputnikovym izmereniyam infrakrasnoi radiatsii (Some Results of Determining Cloud-Top Heights from Satellite-Measured Infrared Radiation).— Trudy Gidrometeorologicheskogo Tsentra SSSR, No.73. 1971.

2. Novosel'tsev, E.P. O stepeni chernoty oblakov verkhnego i srednego yarusov (On the Blackness of High and Medium-Level Clouds).— Trudy GGO, No.196. 1966.

3. Rukovodstvo po meteorologicheskomu ispol'zovaniyu MRL-1 (Instructions on the Use of the MRL-1 Meteorological Radar for Meteorological Purposes).— Leningrad, Gidrometeoizdat. 1968.

4. Valovcin, F.R. Infrared Measurements of Jet-Stream Cirrus.— J. Appl. Meteor., Vol.7, pp.817—826. 1968.

UDC 551.509

UTILIZATION OF SATELLITE RADIATION MEASUREMENTS IN ANALYZING THE TEMPERATURE NEAR THE GROUND

V. G. Boldyrev, V. I. Khamarin

The problem of determining the temperature near the ground from a satellite is discussed. Certain possibilities are considered of utilizing measurements of the actual radiation of the ground (at atmospheric windows) to objectively analyze the ground temperature field. Estimates are presented of errors in temperature interpolation and extrapolation for various types of surfaces.

As is known /5/, it is possible in principle to use infrared and radio-wave radiation measurements of the terrestrial surface to determine its temperature. This possibility is based on the existence of spectral regions over which terrestrial radiation is hardly absorbed by any atmospheric components.

From the practical point of view remote determination of the temperature of a radiating surface (from a satellite, for example) is not at all a simple problem. It is necessary to know the emissivity of the surface area under observation and to be able to take into account the (even if not too large, but still) perceptible effect of the atmosphere on the radiative transfer. It should also be remembered that the temperature must be measured with a high degree of accuracy. It suffices, for example, to recall that the ocean surface temperature must be determined to within less than $1°K$ (possibly even less than $0.5°K$).

At the same time available data on the emissivity of natural surfaces are as yet insufficient (with the exception of the ocean surface, for which the infrared emissivity is very close to unity). The transformation of terrestrial radiation when passing through the atmosphere has not been sufficiently studied (both theoretically and experimentally). For the above reasons it is at present impossible to assume that the temperature of a given area of the earth's surface can be determined from a satellite with an accuracy comparable with that of direct ground measurements.

At present, the most optimistic estimates of the accuracy in determining the surface temperature (by interpreting actual satellite observations) are those presented in /4, 13/.

Smith et al. /13/ explored the possibility of determining the sea-surface temperature from Nimbus satellite 3.8μ window radiation data. The resolving power of water at this wavelength is almost unity (0.98 to be more precise /12/), while the difference between the radiation temperature of the radiation and the actual temperature of the radiating body is relatively small. According to calculations /13/, even in tropical atmospheres with their high water-vapor content (4.7 cm of precipitable water) the difference between temperatures in the nadir angle range of $0-50°$ does not exceed 4°K. In /13/, allowance for this quantity is combined with an estimate of the space variability of measurements (over elementary areas including from 200 to 1000 individual measurements). The distribution curve constructed for such an elementary area enabled any possible effect of cloudiness in the instrument's field of view, as well as the effect of random measurement errors to be eliminated to some extent. The data interpretation technique suggested in /13/ is purely empirical. Comparison of satellite- and ship-measured data on the ocean temperature yields an rms difference of 1.71°K. Since this quantity includes, together with errors of satellite observations, also ship-measurement errors, it is assumed in /13/ that the relative discrepancies can be up to 1°K.

Data of measurements over the $10-11\text{-}\mu$ range carried out from the Cosmos-320 satellite /4/ also show that the mean differences in the ground and radiation temperatures is $3-6°$K. Variations in these differences were found on the average to be 2°K. The inferred surface temperatures in cloud-free regions generally coincide with data of ground meteorological stations to within $2-3°$K. It should be noted that the spatial resolution of the instrument referred to the ground was 10×20 km, and the results were inferred from single measurements. The random error in measuring the radiation temperature was 2°K. Hence the accuracy in determining the ground temperature should be improved by processing some given number of measurements obtained from a sufficiently uniform area.

Available data indicate that correctly organized satellite experiments should yield the surface temperature (primarily the temperatures of sea and ocean surfaces) to within about 2°K (and possibly even more precisely). Such an accuracy is clearly insufficient for local estimates, but may be acceptable in constructing smoothed maps of ocean temperatures from measurements over several days.

In this connection the following points should be clarified: 1) the extent of reducing satellite measurement errors by space filtration of satellite measurements proper and by utilization of ground-network data; 2) the possibility of space interpolation and extra-polation of observation data in order to obtain information in regions for which data are unavailable (due, for example, to the presence of a quasisteady cloud cover).

These problems can be resolved by various smoothing and inter-polation techniques. In this paper we describe some experiments using the optimum interpolation method /2/, which allows to the greatest possible extent for the statistical structure of the quantity under study and the random nature of measurement errors. In addition, this method is most suitable in the case of a sparse measurement network.

Starting with the above results of satellite measurements, one should first consider the possibility of utilizing satellite data to construct maps of sea-surface temperatures. Before passing to this problem we consider briefly data on the statistical structure of the temperature of water, as well as of the ground surface temperature.

TABLE 1. Temperature correlation functions μ_1 and μ_2

Distance (km)	European USSR		North Atlantic (summer)
	winter	summer	
200	0.75	0.68	0.43
400	0.66	0.56	0.35
600	0.56	0.48	0.26
800	0.50	0.39	0.25
1000	0.43	0.30	0.08
1200	0.40	0.23	0.02
1400	0.39	0.17	0.01
1600	—	0.13	0.00
1800	—	0.10	0.00
2000	—	0.10	—

Studies of this structure /1, 3, 9/ indicated that the temperature field is essentially of a small-scale nature. Table 1 lists correla-tion functions μ_2 of the temperature for the Atlantic Ocean and μ_1 for European USSR. It is seen that there is virtually no correlation in the temperature field starting at such relatively small distances as 1000 km. This is confirmed also by Table 2, which gives the matrix of temperature correlation coefficients for weather ships in the North Atlantic. Only infrequently do the correlation coefficients for distances in excess of 1000 km differ significantly from zero. Another characteristic feature is damping of correlation functions in the 0–200 km range /10/. These features put substantial limitations on temperature interpolation and extrapolation.

TABLE 2. Temperature correlation matrix according to weather ships in the North Atlantic in summer

	Ship position		1	2	3	4	5	6	7	8
1	36° N	48° W	1.00 / 0	0.44 / 1031	0.04 / 2184	0.21 / 2021	0.00 / 2943	0.18 / 3194	0.45 / 2805	—0.03 / 4398
2	44	41		1.00 / 0	0.28 / 1510	0.36 / 1018	0.10 / 1999	0.19 / 2186	0.50 / 1806	0.06 / 3420
3	56	51			1.00 / 0	0.28 / 1073	0.50 / 1170	0.27 / 1894	0.11 / 2000	0.30 / 2844
4	52	36				1.00 / 0	0.10 / 1012	0.15 / 1225	0.35 / 1039	—0.01 / 2448
5	62	33					1.00 / 0	0.22 / 831	0.05 / 1241	0.24 / 1687
6	59	19						1.00 / 0	0.26 / 650	0.11 / 1269
7	53	20							1.00 / 0	0.05 / 1820
8	66	01° E								1.00 / 0

Note. The numerator represents the correlation coefficient; the denominator lists the distance between stations.

Consider now numerical experiments in studying the temperature field by the optimum interpolation method. It will be assumed, as usual, that the criterion of success of the analysis is quantity ε, which is the ratio of the standard error (for example, of the inter-polation error) to the variance of the quantity under study. The value of ε depends on the nature of the correlation function of the quantity under study, the number of measurements used in the analysis, and the measuring error.

The specific features of satellite measurements require, in the first place, consideration of the dependence of ε on the measuring error. The latter is denoted by η, which is the ratio of the standard error of measurements to the variance of the quantity being measured. When studying meteorological-network data quantity η is usually taken as constant. As noted above, errors in satellite measurements may differ from those of ground observations. In addition, errors in satellite measurements can be correlated with true temperature values, as well as with the observations themselves. However, it will be assumed here that no correlation exists between the errors in measurements and the quantity being measured.

Consider the effect of an inhomogeneity in the random-error field on the value of ε. In view of the assumptions made, the random error only affects the value of the quantity's variance. The curve $\varepsilon(\eta)$ with η identical for all data-supplying stations was obtained in /6/ up to $\eta = 0.1$. Values of ε and the absolute interpolation error ΔT, obtained for European USSR with an average density of 10 stations over a circle 2000 km in diameter, are as follows:

η	0.00	0 02	0.04	0.06	0.08	0.10	0.20	0.30	0.40
ε	0.18	0.19	0.20	0.21	0.22	0.23	0.26	0.28	0.30
$\Delta T°$	3.06	3.14	3.18	3.30	3.38	3.46	3.68	3.82	3.95

Even for quite substantial measuring errors (encountered sometimes in interpolation of satellite measurements) the interpolation error does not increase too perceptibly. A fact of much greater importance is that the correlation functions of the surface temperature field (those used in /6/, as well as in the present study) are damped out quite rapidly, in conjunction with which the absolute errors in interpolation, even in the ideal case $(\eta = 0)$, are impermissibly high. It can hence be expected that interpolation of data on the ground or radiation temperature even for distances of about 200 km is unsuitable in practice.

It appears slightly more promising to use space interpolation for small distances (up to 100 km) and to filter out measurement "noise" obtained for a sufficiently dense regular (or quasiregular) network of points. An example of such measurements are actinometric and infrared-equipment observation data of the Meteor-series satellites. Data measured by actinometric equipment are obtained, for example, over areas of 50×50 km, the distances between centers of these areas being 50—100 km in the direction parallel to this plane (here consideration is given only to measurements in the $\pm 50°$ nadir angle from the vertical).

It can be assumed in the first approximation that window radiation measurements in the atmosphere pertaining to cloud-free sections can be carried out using corresponding correlation functions for the ground temperature. Variations in the humidity field or in the content of atmospheric ozone probably have little effect on variations in the field of radiation temperatures on a smaller-than-synoptic scale.

To study the possibility of filtration and interpolation of the aforementioned satellite measurements, we examine a simple one-dimensional case of linear filtration, treating both the measurements and the measurement errors as steady random processes.

The temperature correlation functions are approximated in the form

$$\mu_1(\rho) = 0.9e^{-0.08\rho} + 0.1\delta(\rho),$$
$$\mu_2(\rho) = 0.6e^{-0.15\rho} + 0.4\delta(\rho).$$

The first function pertains to European USSR and the second to the North Atlantic (Table 1).

Correlation functions of measurement errors were specified in the form of rapidly damping exponents, corresponding to quite poor correlation of the errors:

$$\mu_3(\rho) = De^{-4\rho}.$$

Here the measurement variance was taken equal to unity, while the variance of measuring errors $(D < 1)$ was specified variously. It was assumed that the errors are not correlated with the measurements.

Table 3 lists the weighting functions and relative errors of filtration and extrapolation of ε for different cases. We now consider filtration $(R = 0)$ and extrapolation to distances of $R = 100$ km and $R = 200$ km for functions $\mu_1(\rho)$ and $\mu_2(\rho)$ when $\mu_3(\rho) = 0.02\ e^{-4\rho}$(cases 1 and 2) and function $\mu_2(\rho)$ when $\mu_3(\rho) = 0.4e$ (case 3).

Table 3 indicates that observations carried out at distances greater than 100 km from the point of filtration virtually do not affect the quality of the latter. Hence in reducing actinometric data it suffices to carry out smoothing within the limits of a single scanning line, while in processing data on infrared observations (with a resolution of 15 × 15 km at the nadir) a square of approximately 100 × 100 km suffices. This is precisely the size of the area selected for operative reduction and automatic decoding of infrared images /7/. Extrapolation of measurement data over distances in excess of 100 km is meaningless for measurements above oceans. The land distance for which there is still sense in extrapolating observations is somewhat greater, but does not exceed 200 km.

TABLE 3. Weighting functions and interpolation errors for various cases

Distance (km)	Weighting function			Interpolation error		
	Case					
	1	2	3	1	2	3
			$R=0$			
0	1.122	1.126	0.978	0.019	0.019	0.240
25	0.128	0.035	0.267			
50	0.074	0.020	0.183			
75	0.046	0.014	0.137			
100	0.031	0.009	0.109			
200	0.008	0.004	0.054			
300	0.002	0.002	0.029			
			$R=100$			
0	1.035	0.641	0.510	0.290	0.626	0.690
25	0.472	0.334	0.310			
50	0.340	0.273	0.255			
75	0.248	0.223	0.215			
100	0.180	0.184	0.183			
200	0.051	0.083	0.097			
300	0.015	0.038	0.052			
			$R=200$			
0	0.956	0.538	0.508	0.390	0.731	0.780
25	0.454	0.304	0.299			
50	0.328	0.249	0.240			
75	0.239	0.203	0.197			
100	0.173	0.168	0.165			
200	0.049	0.076	0.086			
300	0.015	0.034	0.046			

As mentioned above, the magnitude of the measurement error has relatively little effect on the quality of filtration and extrapolation. In any case, for the errors involved in inferring the ocean-temperature field (corresponding to approximately $\eta = 0.2$), obtained in /1, 2/, the satellite measurements carry useful information on the temperature field and supplements climatic data. However, for qualitative analysis of the ground temperature field one should strive to obtain contiguous satellite information with no local blank spaces.

In general, all the above applies only when an individual item of satellite information pertains to an area which is small compared with the distances under study ("point" measurements). This condition is not satisfied in the case of actinometric data. The correlation function for quantities determined for areas of 50×50 km may behave differently than, say, function $\mu_2(\rho)$ for ocean conditions.

Preliminary calculations of correlation functions of radiation temperatures in the $8-12$-μ range using Meteor-satellite measurements show, however, that even on the average (for different meteorological conditions) these functions behave approximately in the same manner as function $\mu_2(\rho)$ /11/.

It follows from calculations that when using infrared measurements it is virtually impossible to eliminate the effect of cloud by space interpolation or extrapolation of radiation-temperature measurements obtained over cloud-free segments. Consequently, the only possibility of obtaining a temperature map at the ground or sea surface without large blank spaces is to combine observations for different dates.

If one constructs maps for some period (for example, three days), then the number of cloud-free sections should increase. This is due to the substantial time variability in the cloud field (as early as after the first 24-hour period the time correlation function for the cloud amount is only 0.2 /8/). The number of available weather ship observations increases correspondingly, and this makes it possible to analyze them simultaneously with satellite data. In any case the density of the observation network (both ground and satellite) should be such that the distance between the node for which interpolation or extrapolation is carried out and the station does not exceed 100 km (for ocean regions).

TABLE 4. Variation in the theoretical interpolation error ε and variation η in the random error for functions $\mu_1(\rho)$ and $\mu_2(\rho)$

	ρ (km)							
	100	200	300	400	500	600	700	800
$\mu_1(\rho)$								
$\eta = 0.02$	0.29	0.36	0.41	0.45	0.47	0.49	0.50	0.51
$\eta = 0.10$	0.32	0.38	0.43	0.46	0.48	0.50	0.51	0.52
$\eta_1 = 0.04$	0.30	0.36	0.41	0.45	0.47	0.49	0.50	0.51
$\eta_1 = 0.20$	0.35	0.40	0.44	0.46	0.48	0.50	0.50	0.51
$\mu_2(\rho)$								
$\eta = 0.02$	0.70	0.76	0.80	0.82	0.84	0.85	0.86	0.86
$\eta = 0.10$	0.72	0.77	0.80	0.83	0.84	0.85	0.86	0.86
$\eta_1 = 0.04$	0.71	0.76	0.80	0.82	0.84	0.85	0.86	0.86
$\eta_1 = 0.20$	0.73	0.78	0.81	0.83	0.84	0.85	0.86	0.86

Note. η_1 equal to 0.04 and 0.20 corresponds to the error of the mobile station, while at the stationary stations $\eta = 0.02$; η equal to 0.02 and 0.10 is the error at every station.

Some confirmation of the above is also provided by the data of Table 4, which lists the results of numerical experiments using the optimum interpolation scheme. Table 4 was constructed on the basis of the following model. The stations $(n = 8)$ were situated symmetrically on a circle with radius ρ about the node, situated at the center of the circle. Note that, according to Table 5, the value of ε does not decrease perceptibly upon further increase in the number of stations. In the measurements, either the same error η was ascribed to all the stations or one station was ascribed an error η_1 different from η. Then, this station (with error η or η_1) was moved along the node with the location of other stations remaining as before. The moving station simulated satellite measurements, while stationary stations simulated measurements of the ground meteorological network. Computational results for correlation functions $\mu_1(\rho)$ and $\mu_2(\rho)$ are listed in Table 4.

TABLE 5. Theoretical interpolation error ε as a function of the number of stations n

n	ρ km							
	100	200	300	400	500	600	700	800
1	0.32	0.42	0.51	0.58	0.64	0.70	0.74	0.78
8	0.17	0.22	0.27	0.32	0.37	0.42	0.46	0.51
12	0.16	0.22	0.27	0.32	0.36	0.41	0.46	0.50

Table 4 shows that, for conditions when function $\mu_1(\rho)$ applies, the success of interpolation for a sparse station network increases perceptibly when an observation appears in the vicinity of the interpolation node. For example, for "ground" stations located 800 km from the node, even when the accuracy is high $(\eta = 0.02)$ the interpolation error is 0.51. The use of the "satellite" measurement at 100 km from the node reduces the interpolation error to 0.35, even if the quality of this measurement is poor $(\eta = 0.4)$. This improvement also applies in the case of temperature fields under oceanic conditions. However, in this case (as mentioned previously) it does not appear possible to carry out interpolation owing to the quite sharp damping of the correlation. In fact the value of the interpolation error ε, even when the distance between the node and the station is only 100 km, is not less than 0.70.

On the whole it can be claimed that satellite observations, even when the random observation error is relatively high, yield improved analysis of temperature near the ground. This is especially true in the case of a sparse station network. Practically, however, correlation in the ground field temperature damps out so rapidly with distance (particularly for oceans) that there is no sense in interpolating and extrapolating observations. The only thing that can be done is to smooth (filter) observations obtained at nodes of a quasiregular network with a very dense grid (node-to-node distances of the order of tens of kilometers). Hence maps of the ground and sea-surface temperature from infrared satellite measurements should be constructed by combining all the observations over the given region during some time interval. When using observations at the applicable radio wavelengths (where the clouds to not interfere with measurements), a map for a specific date can also be obtained.

BIBLIOGRAPHY

1. Boldyrev, V.G. and V.I. Khamarin. K voprosu ob odnorodnosti i izotropnosti polya prizemnoi temperatury (Uniformity and Isotropicity of the Surface Temperature Field).— Trudy Gidrometeorologicheskogo Tsentra SSSR, No.11. 1967.
2. Gandin, L.S. Ob"ektivnyi analiz meteorologicheskikh polei (Objective Analysis of Meteorological Fields).— Leningrad, Gidrometeoizdat. 1968. [English translation by Israel Program for Scientific Translations, Jerusalem. TT 65-50007, Catalog No.1373.]
3. Gandin, L.S., R.L. Kagan, and V.P. Tarakanova. K voprosu o ratsional'nom planirovanii seti nablyudenii za temperaturoi vozdukha (Proper Planning of an Air-Temperature Observation Network).— Trudy GGO, No.228. 1968.
4. Gorodetskii, A.K. Eksperimental'noe issledovanie sobstvennogo izlucheniya Zemli v "okne prozrachnosti" 8—12 mkm so sputnikov (Experimental Study of the Self-Radiation of the Earth in the 8—12 Micron Window from Satellites).— Author's abstract of dissertation. Moscow. 1970.
5. Kondrat'ev, K.Ya. Metodika vvedeniya popravok, uchityvayushchikh vliyanie tolshchi atmosfery pri opredelenii temperatury podstilayushchei poverkhnosti so sputnikov (Technique for Introducing Corrections for the Thickness of the Atmosphere in Satellite Measurements of the Ground Surface Temperature).— Meteorologiya i Gidrologiya, No.2. 1969.
6. Meleshko, V.P. and A.E. Prigodich. Ob"ektivnyi analiz vlazhnosti i temperatury (Objective Analysis of Humidity and Temperature).— Trudy simpoziuma po chislennym metodam prognoza pogody. Leningrad, Gidrometeoizdat. 1964.
7. Sonechkin, D.M. and L.M. Soskin. Automatic Analysis of Cloud Cover by Infrared Photography of the Earth from Meteor Satellites.— Present collection.
8. Sonechkin, D.M. and I.S. Khandurova. Rezul'taty issledovaniya prostranstvenno-vremennoi izmenchivosti oblachnosti na Evropeiskoi territorii SSSR (Results of a Study of the Space-Time Variability of Cloudiness Over European USSR).— Trudy Gidrometeorologicheskogo Tsentra SSSR, No.50. 1969.
9. Khatamkulov, G.Kh. Ob ispol'zovanii nazemnoi informatsii pri ob"ektivnom analize aerologicheskikh polei (Utilization of Ground Information in Objective Analysis of Aerological Fields).— Trudy GGO, No.208. 1967.
10. Hrda, J. O statisticheskoi strukture prizemnogo polya temperatury vozdukha na territorii Chekhii i Moravii (Statistical Structure of Atmospheric Temperature Field near the Ground in Bohemia and Moravia).— Időjárás, Budapest, Vol.72, pp.210—215. 1968.
11. Boldirev, V.G. The Results of Studies of Outgoing Infrared Radiation.— Proc. Inter-Regional Seminar on the Interpretation of Meteorological Satellite Data. Melbourne, Australia. 1968.
12. Buettner, K.J.K. Possible Measurements of Surface Characteristics with Remote Sensors.— XIII COSPAR meeting, Leningrad. 1970.
13. Smith, W.L., P.K. Rao, R. Koffler, and W.R. Curtis. The Determination of Sea-Surface Temperature from Satellite High Resolution Infrared Window Radiation Measurements.— Monthly Weather Review, Vol.98, pp.604—611. 1970.

UDC 551.509

AUTOMATIC ANALYSIS OF CLOUD COVER BY INFRARED PHOTOGRAPHY OF THE EARTH FROM METEOR SATELLITES

D. M. Sonechkin, L. M. Soskin

A program is described for computer identification of clouds on infrared photographs of the earth obtained from Meteor-series satellites. An example of a computer cloud chart is given.

In composing a program for the computer analysis of infrared photographs of the ground obtained from Meteor-series satellites, use was made of experience accumulated in experiments with programmed processing of ESSA TV photographs /1/. The purpose of this paper is, first, to clarify the applicability of the method developed in /1/ to low-resolution photographs obtained in another (infrared and invisible) spectral region, and second, having reference to the substantially smaller volume of infrared information, to effect data processing in a reasonable (or close to it) time.

The composition of a program of automatic processing and analysis of infrared photographs supplied from Meteor-series satellites was facilitated by the fact that ready-made programs were available for feeding this information to a Minsk-22 computer and for its geographic referencing. These programs were compiled at the Department of Satellite-Data Analysis of the Hydrometeorological Center of the USSR by V. I. Solov'ev.

The difference between the programs developed for Meteor satellites and for ESSA satellites are due primarily to differences in the geometric structure of obtaining infrared photographs. Instead of frame-by-frame photographs use is made of element-by-element photography with mechanical scanning of the area along the flight path of the satellite over a band approximately 1000 km wide. Automatic processing was carried out with 68 measurements of infrared-emission intensity along each scanning line. This intensity has a latitudinal variation unrelated to the cloud cover structure, and this complicates cloud-identification computation. This variation is eliminated by the use of sliding averages of the measured intensities on a scale commensurable with the dimensions of large-scale cloud formations, i. e., of the order of 1000 km along the satellite flight path. The resulting averages were eliminated from each specific measurement after some nonlinear filtering, which smoothed out the extrema of the average curve. The measurements normed in the above manner served for the primary description of the infrared image which was the object of the analysis. In addition to the measurements proper, the computer memory also stored information on the time when each line of the orbital infrared photograph was taken.

As in the ESSA program, clouds were identified by means of a two-stage procedure. First the orbital band of the infrared photograph was subdivided into local segments of 8 X 8 elements each (8 elements within the photography band), and the type of infrared image characterizing this photograph was determined for each element. In order to speed up processing time and to ensure invariance to shifts and rotations of the image, the coefficients of the expansion of the covariance matrix of the infrared signal in terms of eigenvectors were not used; instead, single- and two-point moments of first and second order were used, characterizing the average level of the normed radiation value, the range of values for individual expansion elements within the segment, and the predominating size and spatial anisotropy of inhomogeneities of the radiation field. The specific form of the characteristics was varied until a satisfactory selection was obtained.

In calculating the identification features it was important to first eliminate from the photograph the signal error, which is inherent in the numerical system of transmission used in the infrared TV system. For this we worked out a special algorithm for the adaptive filtration of the image, consisting in linear interpolation of the signal level for each individual element of the image from its closest neighbors. If the actual value of the signal differed in absolute magnitude from the interpolated value by more than some threshold value, it was assumed that the signal was subjected to interference and then, in subsequent analysis, its interpolated value was used instead of the actual value. If the difference was not greater than the threshold value it was assumed that there was no interference.

Such linear filtration usually results in smoothing out the image and distorting the picture. To avoid this, the threshold value was varied, depending on the structure of the image in the vicinity of the element being analyzed. For this, sliding averages (with exponential infinite memory) of the differences which did not exceed the threshold value were used, while the actual values of the signal were

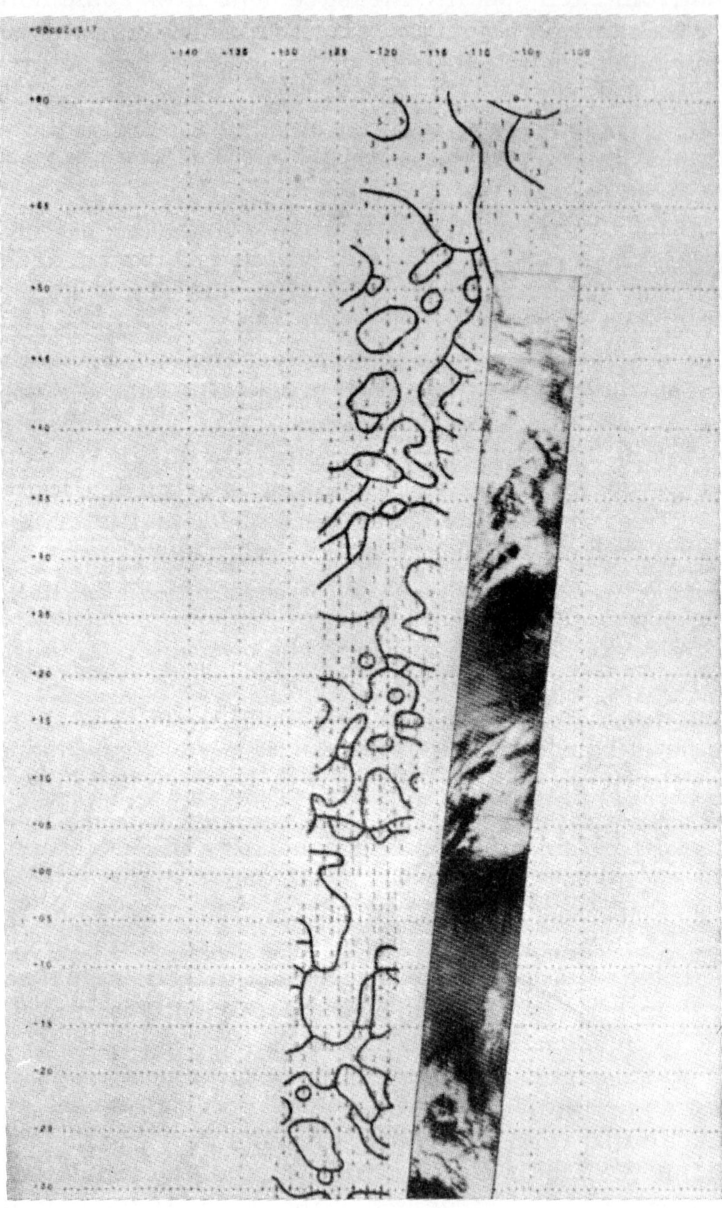

FIGURE 1

compared with its interpolated values.   These differences characterized the structure of the radiation field in the vicinity of the element under study.   Hence in subsequent comparison in regions with (on the average) small variations in the radiation field, an error was defined as a relatively small signal surge, while in regions with relatively small variations only a substantially large surge was regarded as an error.   Here the adaptive filtration algorithm has the form

$$\widehat{B}_{00}[t] = \frac{1}{8}\left(\sum_i \sum_j B_{ij}[t] - B_{00}[t]\right), \quad i = -1, 0.01, \quad j = -1, 0, 1,$$

where $B_{00}[t]$ and $\widehat{B}_{00}[t]$ are the actual and interpolated signals for the picture-expansion element at the t-th step; $B_{ij}[t]$ are the values of the signal for the expansion elements neighboring that under consideration (three in the preceding line, two in the same line as the element under study, three in the subsequent line);

$$\delta B[t] = \left(B_{00}[t] - \widehat{B}_{00}[t]\right),$$

$$|\delta B[t]| - \Delta[t] = \begin{cases} >0, \text{ then } B_{00}[t] = \widehat{B}_{00}[t], \quad \Delta[t+1] = \Delta[t], \\ \leqslant 0, \text{ then } B_{00}[t] = B_{00}[t], \\ \Delta[t+1] = \Delta[t] + k\,|\delta B[t] - \Delta[t], \end{cases}$$

where $\Delta[t]$ is the current threshold value, while k is a constant (smaller than unity), which determines the effective area of the region for which the structure of the radiation field is estimated.

After eliminating errors by means of a self-correcting program described in /1/, we found an accumulation of points (type of infra-red images).

The subsequent work involved calculating a priori, conditional and a posteriori probabilities of various clouds for different types of images, and the nature of the surface was fully analogous to that of the ESSA program.   The difference was that, having reference to the specific features of infrared photography, it was possible to identify six cloud types graduated with respect to their amount and the height of their upper boundary, unlike the three cloud-amount graduations determined by the ESSA program.

It was assumed that the result of the identification represented that type of cloud corresponding to the maximum of the a posteriori probability.   It is determined for each segment of the photograph. The result of identification is presented in the form of a Mercator projection on a scale of 1:30,000,000.   Processing time for a complete infrared-photography orbit is about 2 hours, i. e., processing is carried out in almost actual time (the photography time of this orbit is 1.5 hours). An example of data processing is shown in Figure 1.

BIBLIOGRAPHY

1. Solov'eva, I.S., D.M.Sonechkin, and V.F.Kharitonov. Obrabotka i analiz televizionnykh izobrazhenii oblachnosti na EVM (Computer Processing and Analysis of TV Cloud Pictures).— Trudy Gidrometeorologicheskogo Tsentra SSSR, No.73. 1971.

UDC 551.509

EFFECT OF CLOUD COVER ON THE VARIABILITY OF OUTGOING RADIATION

V. G. Boldyrev, L. I. Koprova

Some results are considered of statistical processing of actinometric data obtained by Meteor satellites. Depending on the cloud situation, we obtained the average latitudinal variation in the radiation temperature (from measurements in the 8–12-μ window) and the intensity of short-wave radiation (from measurements in the 0.3–3-μ band), their rms deviations, as well as empirical curves of radiation temperature distribution. These data provide insight into the limits of variability of the radiation characteristics under study.

Satellite-radiometer measurements of integral short- and long-wave radiation have led to the first global maps of albedo, self-radiation and the radiation balance of the planet Earth. These maps were constructed for different time-averaging periods — from two weeks /2/ to seasons /1/ and several years /3/. Analysis of these maps has made it possible to single out interesting features of the radiation regime on a planetary scale and to relate these features to characteristics of the ground—atmosphere system, in the first place to the cloud field and temperature field of the terrestrial surface.

It is especially clear that it is precisely the variability in the cloud cover which is the principal factor responsible for the relatively short-lived space and time variations in the radiation field of the earth. The cloud cover is also responsible for the fact that the distribution of outgoing radiation intensity, obtained by averaging over various areas, can differ from the standard for certain seasons and geographic regions. Examples of the distribution of radiation temperatures in the 8–12 and 3–30μ regions of radiation temperatures averaged over areas of 1000 $\times$ 1000 km^2 (Pacific, June—August 1969) are presented in /1/. In addition we calculated the corresponding distributions for April—May (spring), September—November 1969 (autumn) and December 1969—February 1970 (winter).

Table 1 was obtained by processing actinometric data obtained by Meteor satellites; the data show that deviations in the statistical distribution of space-averaged values of long-wave radiation from a normal distribution are typical.

TABLE 1. Latitudinal variation in maximum and minimum values of $t(°C)$, B $(cal/cm^2 \cdot min \cdot ster) \cdot 10^{-2}$, and σ_t and σ_B for various situations

Situation number	t min	t max	σ_t min	σ_t max	B min	B max	σ_B min	σ_B max	Number of occurrences
				Zone I					
1	−24.5	−21.4	12.1	14.5					1
2	−27.7	−10.5	4.1	12.0	5.0	18.9	2.2	7.5	5
3 and 4	−16.5	−2.7	6.1	11.9	5.0	18.9	2.2	4.7	7
5	−11.2	−4.6	5.7	9.7					2
6 and 7	−18.7	−1.9	3.1	9.1	5.0	11.4	2.2	7.5	8
				Zone II					
1	−18.8	6.6	9.6	16.9	5.7	8.8	4.4	5.0	7
2	−15.5	11.2	5.2	13.4	4.6	16.6	2.5	5.8	13
3 and 4	−16.7	11.4	2.4	12.9	5	9.1	2.2	3.4	16
5	−8.7	13.2	5.4	14,2	4	17.4	2.6	5.0	7
6	1.6	14.6	2.1	10,4	4	8.3	2.2	4.6	9
7	1.5	18.5	2,2	5,1	4,4	8	3	5.6	4
				Zone III					
1	−8.5	6.6	10.4	21.4	8.8	13.7	5.5	7.6	4
2	−4.5	15.7	6.2	21.8	5.3	17.8	2.2	9.3	12
3 and 4	−4.2	17.4	2	12.5	4.1	10.2	2.2	8,4	14
5	−3.6	17.2	6.8	17.1	5.2	13.8	2.3	7.5	7
6	7.0	23.3	2.2	10.2	5.1	8.7	2.4	4.1	21
7	15.4	17.4	2.2	3.3	5.0	9.7	3.4	3.8	4
				Zone IV					
1	−12.4	15.7.	11.4	17.8	6.4	12.2	4.1	7.3	4
2	0.1	15.8	15.6	21.4	4	12.9	3.3	8.0	12
3 and 4	0.7	18.0	3.8	14.8	4.6	13.4	2.2	7.0	17
5	−2.0	28.3	3.6	20.1	4.6	16.1	2.2	7.2	33
6	9.0	25.0	2.4	10.6	5.0	12.4	2.2	8.8	26
7	17.0	24.3	1.2	4.4	5.4	7.5	2.2	3.9	8
				Zone V					
1	−11.0	16.1	3 4	13.3	5	23.4	3.3	10.1	6
2	−7.9	11.6	12.7	19.7	5	19.8	5.6	11.0	6
3 and 4	−9.8	16.3	2.8	10.4	4.3	7.9	2.3	7.9	12
5	−8.0	20.1	3 6	16.6	5.1	18.6	3.0	7.4	14
6	2.6	23.2	2.0	9.7	5.6	11.4	2.3	6.6	19
7	13.9	22.2	2.2	3.6	5.0	10.3	2.5	4.0	6
				Zone VI					
1	—	—	—	—	—	—	—	—	—
2	−20.5	13.2	4.2	18.0	5.0	23.3	3.0	11.0	21
3 and 4	−15.3	13.6	3.0	13.7	4.8	23.9	2.2	7.4	42
5	−19.0	16.1	3.2	17.8	5.9	24.6	2.2	8.8	40
6	−4.0	17.6	2.2	13.5	5.8	22.6	3.2	9.6	16
				Zone VII					
8	−42.8	−0.9	2.8	13.6	5	34.2	2.2	6.5	93
Zone number				Situations 1 − 8					
I	−27.7	−1.9	3.1	14.5	5.0	18.9	2.2	7.5	23
II	−18.8	18.5	2.3	16.9	4.0	17.4	2.2	5 7	56
III	−8.5	23.3	2.2	21.8	4.1	17.8	2.2	9.3	62
IV	−12.4	28.3	1.2	21.4	4.0	16.1	2.2	8.8	100
V	−11.0	23.2	2.2	19.7	4.3	23.4	2.2	11.0	63
VI	−20.5	17.6	2 3	18.0	4.8	24.6	2.2	11.0	119
VII	−42.8	−0.9	2.8	13.6	5.0	34.2	2.2	6.5	93

In a number of regions the statistical radiation distribution is even bimodal and approaches rectangular. All this makes it difficult to estimate the accuracy of climatological quantities which are obtained and requires more detailed study of the nature of radiation variations due to variations in the cloud cover. Here, in the first place, it is expedient to obtain the average values of the variability in the radiation and its values as a function of the nature of the cloud cover in a given region and during different seasons. This very work-consuming and complex problem is not treated here.

This article deals only with very preliminary estimates of the statistical distribution of radiation as a function of the nature of the cloud cover, obtained by processing actinometric observations from Meteor satellites (primarily measurements of radiation temperature in the $8-12$-μ region and somewhat of the intensity of reflected short-wave radiation in the $0.3-3$-μ region).

Data obtained with an average resolution radiometer (about 50×50 km at the nadir) do not allow one to study the relationship between mesoscale variations in radiation and the cloud cover, because the distance between scanning lines (when the distance from the subsatellite points is not too excessive) is about 120 km while the distance between the centers of the viewing field of the radiometer along the scanning line in the vicinity of the nadir is about 50 km. Hence such data yield statistical characteristics pertaining to quite large zones — about $500-700$ km perpendicular to the direction of the orbital projection and $1000-2000$ km along the orbit. The dimensions of these zones are commensurable with those of synoptic-scale cloud formations.

If areas of such a size (characterized by the presence of a given cloud form) of some typical cloud formation or some kind of surface are selected, it is possible to obtain data on the range of variation in radiation intensities averaged over these areas and on the nature of the variation in these averaged quantities. These data can be compared with characteristics of the cloud cover and of the surface. By processing satellite measurements, the intensities of short-wave radiation and the radiation temperatures were averaged over an area of 500×1800 km from 225 individual measurements. For these areas we also calculated the variances and various other statistical characteristics of the radiation temperature in the $8-12$-μ region and intensities of short-wave radiation in the $0.3-3$-μ region. The number of measurements within a single segment was thus approximately commensurable with the number of measurements subjected to averaging in the construction of seasonal radiation maps /1/.

Processed data comprised radiation measurements obtained from more than one-hundred orbits of Meteor-series satellites for the April—December 1969 period. The total number of nonoverlapping (regarded as independent) segments was more than 500. We examined a part of the Pacific Ocean extending from 50°N to 50°S, the eastern part of North America, and the Antarctic. Seven zones were selected:

I 61—77°N and 115—160°W in North America
II 31—58°N and 125—155°W in the Pacific
III 11—30°N and 125—152°W in the Pacific
IV 10°S—10°N and 120—150°W in the Pacific
V 11—30°S and 120—150°W in the Pacific
VI 31—73°S and 105—145°W in the Pacific
VII 68—82°S and 90°E—165°W in the Antarctic

These zones form a kind of meridional section along a narrow range of longitudes. Each of these zones has features of its own in the cloud-cover distribution. Data of measurements for each latitudinal band were subdivided into groups in accordance with a given situation. We considered the following eight (obviously quite arbitrary) situations: 1) heavy cumulonimbus, frontal cloudiness; 2) cumulus and stratocumulus, occupying about 75% of the segment under study; 3) cumulus (cellular); 4) stratus occupying more than 75% of the segment under study; 5) moderate cloudiness with a sharply delineated boundary within the segment; 6) moderate cloudiness without a sharply delineated boundary within the segment; 7) clear (ocean); 8) ice, snow, cloudiness (without specification of form).

In order to refer each radiation-temperature and short-wave radiation reading to some situation, it was necessary to use additional satellite information in the form of infrared and TV photographs obtained simultaneously with radiation measurements.

The above method was used to subdivide the data into groups corresponding to geographic regions and cloud situations, in each of which were selected maximum and minimum values of the mean radiation temperature t and short-wave radiation intensity B, as well well as of the rms deviations σ_t and σ_B.

Table 1 shows the latitudinal variation in these quantities without breakup into situations. Maximum and minimum values of the space-averaged radiation temperatures have a clearly expressed characteristic latitudinal variation. The rms deviations of the radiation temperature have virtually no latitudinal variation. The table also lists the latitudinal variation in t, B, σ_t and σ_B for the different situations and gives the number of independent

TABLE 2. Empirical distribution of radiation temperature and of short-wave radiation intensity for various situations

Situation number		1	1	2	3	4	5	6	7	8
Orbit number . .		542	659	161	615	542	102	190	307	190
Zone number . .		III	II	III	VI	VI	III	III	IV	VII
$t°$ C										
from	to									
—60	—55									
—55	—50	0								
—50	—45	0.4								
—45	—40	3.1		0						0
—40	—35	4.9	0	0.4						10.2
—35	—30	4.0	8.5	0						16.9
—30	—25	2.6	17.0	0.4	0	0				15.3
—25	—20	2.2	13.9	3.1	2.3	0.9				38.4
—20	—15	1.7	6.7	3.6	3.8	1.9				17.9
—15	—10	1.3	11.2	6.7	12.4	10.5				1.0
—10	—5	2.6	6.7	10.3	15.3	23.5		0	0	0
—5	0	2.6	5.8	10.3	16.2	29.8		0.6		
0	5	3.5	3.5	9.0	23.9	25.0	0	0.6		
5	10	16.5	15.2	7.6	25.8	8.1	0.4	2.6		
10	15	6.7	11.2	11.7	0	0	29.7	62.0	0	
15	20	6.2	0	30.6			47.1	34.0	9.7	
20	25	35.8		5.8			22.6	0	90.2	
25	30	4.9		0			0		0	
30	35	0								
35	40									
$B \cdot 10^{-2}$ cal/cm$^2 \cdot$min. $\cdot$ster										
from	to									
0.02	0.04	0		0		0	0	0	0	6.9
0.04	0.06	0	0	13.7	0	4.4	33.7	42.6	14.2	7.7
0.06	0.08	29.6	23.6	17.7	12.6	8.5	25.7	19.3	32.8	41.0
0.08	0.10	12.0	22.1	19.5	16.3	9.8	18.6	11.3	36.0	9.3
0.10	0.12	7.1	29.0	11.1	12.1	14.7	7.1	5.3	10.2	7.7
0.12	0.14	8.2	22.1	9.7	10.5	14.7	7.5	6.0	4.0	24.0
0.14	0.16	4.3	3.0	6.2	11.0	14.3	4.4	4.0	2.2	2.3
0.16	0.18	3.2	0	8.0	7.8	14.2	2.6	6.0	0.4	0.7
0.18	0.20	4.9		6.6	4.7	8.4	0	2.0	0	0
0.20	0.22	6.5		4.0	6.8	3.1		1.3		
0.22	0.24	4.3		1.7	4.2	0.4		0		
0.24	0.26	5.4		0.4	4.7	0.4				
0.26	0.28	8.7		0.8	5.2	0				
0.28	0.30	2.7		0	2.6					
0.30	0.32	1.0			1.0					
0.32	0.34	0.5			0					
0.34	0.36	0								
0.36	0.38	0.5								
0.38	0.40	0								
0.40	0.42									

174   ADVANCES IN SATELLITE METEOROLOGY

TABLE 3. Average empirical distributions of radiation temperature for various latitude zones and situations

Situation number.		1—2						6—7					
Zone number …		I	II	III	IV	V	VI	I	II	III	IV	V	VI
t° C from	to												
—60	—55												
—55	—50			0	0	0							
—50	—45			0.1	0.4	0.2	0		0				
—45	—40	0	0	0.8	0.4	0.3	0.6		0.2	0	0		
—40	—35	2.4	0.5	1.2	0.8	0.7	1.5	0	0.9	0.3	0.2	0	
—35	—30	7.6	3.1	1.6	1.3	1.5	2.2	0	3.6	0.8	0.4	0.1	0
—30	—25	11.5	5.6	1.1	2.5	2.4	4.8	2.5	6.3	0.7	0.4	0.3	0.5
—25	—20	16.5	7.0	2.4	2.4	3.9	6.7	8.2	5.0	1.2	1.0	5.4	3.5
—20	—15	11.5	5.4	2.9	2.7	5.1	5.4	14.4	7.3	1.5	2.9	11.6	7.0
—15	—10	9.9	9.0	2.4	3.1	6.1	6.4	29.3	8.7	2.3	2.6	1.6	13.6
—10	—5	9.7	7.0	4.0	2.4	7.2	9.4	14.8	12.5	3.5	4.5	1.9	18.6
—5	0	22.2	8.1	3.8	6.5	9.0	14.1	17.7	13.9	4.1	5.5	4.2	23.1
0	5	6.2	7.1	3.8	6.6	11.9	22.8	8.3	18.3	3.9	6.1	7.7	17.5
5	10	3.0	18.9	7.8	7.5	13.5	17.3	5.9	19.2	8.3	8.0	12.8	13.7
10	15	0.1	20.2	12.2	11.1	12.3	7.4	0	4.1	17.3	10.5	16.8	3.2
15	20	0	6.9	26.3	22.1	10.8	0		0	36.2	22.0	34.1	0.2
20	25		0.9	18.1	20.7	14.0				19.1	25.2	2.9	0
25	30		0	11.6	7.7	0.3				0.5	9.1	0	
30	35			0	2.0	0.2				0.3	1.4		
35	40				0	0				0	0		

Situation number.		6—7						7					8	
Zone number . . .		I	II	III	IV	V	VI	I	II	III	IV	V	VI	VII
$t°$ C from	to													
—60	—55													
—55	—50													0
—50	—45				0									0.1
—45	—40				0.1									1.2
—40	—35		0		0.1								0	3.5
—35	—30		0.1		0.1				0				0.6	5.1
—30	—25	0	3.6		0.1	0	0		0.2				2.2	11.0
—25	—20	0.5	6.8	0	0.2	0.3	0.5		0.2				4.7	16.8
—20	—15	0.7	2.7	0.1	0.3	0.8	1.1		0.2				8.2	20.2
—15	—10	0.5	3.1	0.8	0.5	0.6	3.1		0.0		0		18.9	22.2
—10	—5	0.7	1.9	0.8	0.7	1.2	5.6		0.4		0.1		22.6	13.1
—5	0	0.9	5.5	1.7	1.8	2.4	7.7		0.2		0.2		31.1	7.8
0	5	2.1	9.7	4.9	2.1	5.0	18.0		1.1		0.2	0	10.4	0
5	10	5.4	10.1	9.8	3.6	10.6	30.5		2.0		0.5	0.9	0.8	
10	15	6.0	26.1	29.0	7.6	15.9	16.2		29.3	0	6.2	14.1	0.3	
15	20	8.6	28.9	33.4	28.2	30.1	14.0		63.1	40.6	33.7	31.3	0	
20	25	39.5	0.9	16.6	39.9	27.4	2.5		3.1	49.0	49.1	50.0		
25	30	30.0	0	2.9	15.1	5.3	0		0	10.3	10.4	3.6		
30	35	4.5		0	0.1	0.3				0	0.1	0		
35	40	0.5			0	0					0			

measurements from which the minimum and maximum values of
these quantities were selected. Over the Antarctic it is difficult to
distinguish between the nature of individual situations: snow, ice
and cloud fields merge to produce a gray background on infrared
photographs; in the presence of clouds it is impossible to decipher
the information even on TV photographs.

The following information can be derived from Table 1. Heavy
cumulus and frontal clouds (situations 1 and 2) correspond to the
greatest difference between $t_{\min}$ and $t_{\max}$ (as high as 35°), and also
to the greatest rms deviations, which are of the same order of
magnitudes as the radiation temperatures. The greatest variations
in this temperature ($\sigma_t = 20°$) occur for a considerable cloud cover,
i.e., heavy frontal clouds in moderate latitudes or individual
cumulonimbus and clouds of the intertropical convergence zone in
equatorial latitudes.

The rms deviations for stratus and cellular cumulus (situations
3 and 4) are somewhat smaller. Then, as the cloud amount
decreases (situations 5 through 7), the temperature of the radiating
surface increases and its rms deviation decreases. For cloud-free
conditions (situation 7) the values of the minimum and maximum
radiation temperatures become closer ($t_{\max} = 24°$, $t_{\min} = 18°$ at the
equator) and the rms deviations are small ($\sigma_{\max} = 4.4°$, $\sigma_{\min} = 1.2°$).
In particular, $\sigma_{\min} = 1.2°$ for cloud-free conditions characterizes to
some extent the random error of measurements. It can apparently
be claimed that the random error at above-zero temperatures is
now higher than the above value. It should be noted that the small
number of occurrences of certain situations (Table 1) shows that
these are less frequent than others; however, these data pertain to
clearly defined situations (either truly clear weather, or clearly
expressed cumulus, etc.).

The empirical distribution curves of t and B pertaining to
different situations differ markedly in shape. Examples of empirical distributions (in percents), corresponding to seven cloud-cover
situations, are listed in Table 2.

Heavy cumulus and frontal clouds have multimodal radiation-
temperature and short-wave radiation intensity distributions. For
cloud-free conditions, when the instrument views a more uniform
field, t and B have a narrow single-mode distribution. The
empirical distribution for situation 1 is given for the case with
maximum rms deviation of the radiation temperature ($\sigma_t = 21.8°$).
The temperature and radiation-intensity distributions for stratus
become almost symmetrical and possess one mode. Finally, for
clear skies or for scattered stratus the variability in the radiation
(particularly long-wave) becomes quite small.

Table 3 lists averaged empirical distributions of radiation temperature for various latitude zone and cloud situations. The principal features of distributions typical of the different situations (see Table 2) persist also in the average empirical distributions.

The empirical distributions for situations 1 and 2 are wide and multimodal. The latitudinal variation in empirical distributions is clear: their principal maxima for equatorial zone IV and near-equatorial zones III and V virtually coincide; for northern zone I and southern zone VI the empirical distributions are close and shifted in the direction of lower temperatures.

At moderate cloudiness (situations 6 and 7) the averaged empirical distributions become narrower and have virtually no latitudinal variation. For cloud-free weather (situation 7) the distributions are narrow and have one mode, and do not depend on the latitude. Situation 7 is not generally observed in zones I and VI.

Situation 8 is characteristic for zones VI (Amundsen Sea, Antarctic shore) and VII (Antarctic). Here the distributions are, on the average, wide and the empirical distribution for zone VII is shifted toward lower temperatures.

The above regularities in the behavior of statistical character- istics of the radiation temperature and short-wave radiation do not suffice to uniquely determine the specific cloud form. These data give only some idea concerning the limits of variability in the short-wave radiation and radiation temperature under a given set of meteorological conditions.

BIBLIOGRAPHY

1. Boldyrev, V.G. and I.P.Vetlov. Prostranstvennaya i vremennaya izmenchivost' ukhodyashchei radiatsii (Space and Time Variability of Outgoing Radiation).— Meteorologiya i Gidrologiya, No.10. 1970.
2. Raschke, E. and W.R.Bandeen. The Radiation Balance of the Planet Earth from Radiation Measurements of the Satellite Nimbus II. —J. Appl. Meteor., Vol.9, pp.215—238. 1970.
3. Vonder Haar, T.H. and V.E.Suomi. Satellite Observations of the Earth's Radiation Budget.— Science, Vol. 163, p.667. 1969.

UDC 551.509

STATISTICAL CHARACTERISTICS OF THE TEMPERATURE FIELD NEAR THE GROUND FOR EUROPE

A. V. Karpov, V. I. Khamarin

Some results are considered of a statistical analysis of the spatial structure of the summer and winter temperature field near the ground for Europe. It is shown that homogeneity and isotropicity can be postulated for the field of centered and normalized temperatures in individual climatic regions.

As a rule, the structure of the temperature field near the ground is analyzed on the assumption that deviations from the mean are homogeneous and isotropic /3, 4/. However, individual calculations of space correlation functions of the temperature show that sometimes substantial departures from homogeneity and isotropicity may occur /2/. It thus became necessary to study in more detail the structure of the temperature field near the ground, especially as the terrestrial-surface temperature can be measured from artificial satellites. Meteorological satellites make it possible to obtain information on the underlying-surface temperature, which can be used as additional information in refining temperatures near the ground. However, satellite measurements must be matched with the latter because they differ qualitatively from ground observations. One possible method of matching these two fields is the optimum matching method, based on a knowledge of the statistical structure of the fields under study and on their statistical interrelationship. Objective analysis of the temperature field, as well as practical utilization of satellite radiation measurements require that the space structure of the temperature field be known.

The structure of the temperature field near the ground in Europe, during winter, is studied in /2/, where corresponding results for summer, and also the possibility of parametrization of the temperature field near the ground by means of natural functions are examined. Such an approach to the study of the structure can be

implemented by calculating temperature correlation matrices and
their eigenvectors, the determination of which has been discussed
extensively in meteorological literature.

The starting information comprised air temperatures taken
from surface synoptic weather maps. In the region lying approximately between 78 and 35°N and 10°W and 70°E we selected 140
stations (from 1 June 1962 to 10 August 1966), which were more or
less uniformly distributed over the entire area. Data on the air
temperature, taken from one synoptic map at selected points,
represent in our problem one realization of a random process. A
total of 150 such realizations were treated, each 3—5 days removed
from the other. All the data pertain to 0300 hrs Moscow time.

TABLE 1. Temperature covariance matrix (55°N, summer)

Longitude	7°W	1°W	8°E	12°E	20°E	30°E	37°E	44°E	49°E	72°E
7°W	4.49	2.18	13.50	12.60	0.43	0.62	0.97	-0.01	0.28	-0.39
1		7.27	2.34	1.34	-0.08	0.45	0.81	0.49	-0.75	0.20
8°E			4.90	2.96	2.30	1.10	0.92	-0.22	-0.98	-0.60
12				5.53	2.98	1.49	1.04	-0.68	-1.22	-0.34
20					9.60	4.14	3.84	2.43	3.28	0.57
30						9.18	5.36	2.41	2.53	1.53
37							12.60	8.70	7.95	1.34
44								14.80	12.70	2.60
49									18.50	4.90
72										14.20

TABLE 2. Temperature correlation matrix (55°N, summer)

Longitude	7°W	1°W	8°E	12°E	20°E	30°E	37°E	44°E	49°E	72°E
7°W	1	0.38	0.29	0.25	0.07	0.10	0.12	0.00	0.03	-0.05
1		1	0.39	0.21	-0.01	0.06	0.08	0.05	-0.06	0.02
8°E			1	0.59	0.33	0.16	0.12	-0.02	-0.10	-0.07
12				1	0.41	0.21	0.12	-0.07	-0.12	-0.04
20					1	0.44	0.35	0.20	0.24	0.05
30						1	0.50	0.21	0.19	0.13
37							1	0.64	0.52	0.10
44								1	0.77	0.18
49									1	0.30
72										1

The correlation function for a homogeneous and isotropic field
depends only on the interstation distance. The inhomogeneity and
anisotropicity of the temperature field is estimated by comparing

elements of the correlation matrix for different station pairs situated under different geographic conditions. Tables 1 and 2 list the covariance and correlation matrices for stations located at 55°N. For comparison, Tables 3 and 4 give similar matrices for winter /2/.

TABLE 3. Temperature covariance matrix (55°N winter)

Longi-tude	7°W	1°W	8°E	13°E	22°E	30°E	34°E	49°E	65°E	73°E
7°W	23.60	18.90	0.87	-1.33	-8.79	-14.32	-5.58	-9.30	-13.40	-9.42
1		26.80	1.18	1.83	0.34	-6.13	3.51	-0.62	-4.24	-0.26
8°E			19.70	11.10	10.50	12.70	13.70	11.20	-1.23	-0.82
13				13.30	13.90	13.30	15.40	13.80	2.23	3.32
22					40.00	34.40	34.60	28.80	9.00	1.50
30						55.50	43.80	29.80	13.50	3.60
34							60.80	36.60	10.30	1.80
49								71.90	25.80	5.40
65									85.80	70.90
73										145.60

TABLE 4. Temperature correlation matrix (55°N, winter)

Longi-tude	7°W	1°W	8°E	13°E	22°E	30°E	34°E	49°E	65°E	73°E
7°W	1	0.75	0.04	-0.08	-0.28	-0.40	-0.15	-0.23	-0.30	-0.21
1		1	0.05	0.09	0.01	-0.16	0.08	-0.01	-0.12	-0.00
8°E			1	0.69	0.37	0.38	0.40	0.30	-0.03	-0.01
13				1	0.60	0.49	0.54	0.44	0.07	0.09
22					1	0.73	0.70	0.54	0.16	0.05
30						1	0.75	0.46	0.20	0.05
34							1	0.55	0.15	-0.02
49								1	0.34	0.07
65									1	0.64
73										1

Tables 3 and 4 imply that the temperature correlation changes markedly in winter from west to east. For western regions the correlation function (correlation coefficient) passes through zero when the distance between the stations is about 1000 km. For eastern regions of this latitude the correlation is relatively close (at a distance of 1000 km the correlation coefficient is $\mu = 0.5$). In summer (Tables 1 and 2) the correlation coefficients increase smoothly from west to east. On the average, for 55°N, the correlation coefficients in summer are somewhat lower than in winter.

To estimate the inhomogeneity and anisotropicity of the field one must know the correlation coefficients μ for a given distance ρ at different points of the field. We now determine μ for fixed ρ, with a

TABLE 5. Temperature correlation coefficients when $\rho = 200$ km

Coordinates	5°W	0°	5°E	10°E	15°E	20°E	25°E	30°E	35°E	40°E	45°E	50°E	55°E	60°E
Winter														
60° N	0	0.81	0.88	0.86	0.84	0.87	0.90	0.87	0.78	0.80	0.83	0.75	0.75	0.85
55	0.80	0.72	0.75	0.83	0.89	0.90	0.83	0.78	0.93	0.91	0.87	0.84	0.84	0.85
50	0.80	0.69	0.69	0.70	0.72	0.79	0.75	0.74	0.87	0.72	0.85	0.90	0.91	0.90
45		0.90	0.65	0.56	0 60	0.89	0.75	0.53	0.65	0.60	0.64	0.67	0.75	0.77
40	0.80	0.95	0.88	0.60	0.53	0.51	0.79	0.94	0.80	0.83	0.85	0.78	0.69	0.59
Summer														
60° N		0.78	0.80	0.80	0.77	0.70	0.67	0.84	0.83	0.83	0.84	0.84	0.82	
55	0.68	0.80	0.75	0.75	0.80	0.75	0.73	0.76	0.90	0.90	0.89	0.89	0.89	0.88
52				0.82	0.68	0.64	0.73	0.78	0.80	0.80	0.83	0.85	0.84	0.80
45			0.87	0.80	0.74	0.77	0.82	0.83	0.83	0.81	0.74	0.75	0.74	
40		0.70	0.66	0.63	0.62	0.65	0.71	0.80	0.84	0.85	0.86	0.85	0.82	

TABLE 6. Temperature correlation coefficients when $\rho = 400$ km

Coordinates	5°W	0°	5°E	10°E	15°E	20°E	25°E	30°E	35°E	40°E	45°E	50°E	55°E	60°E
					Winter									
60° N		0.57	0.74	0.70	0.67	0.73	0.79	0.70	0.57	0.55	0.64	0.58	0.54	0.54
55	0.40	0.37	0.52	0.65	0.75	0.78	0.73	0.69	0.84	0.78	0.72	0.70	0.70	0.72
50	0.65	0.38	0.47	0.49	0.51	0.63	0.53	0.50	0.56	0.70	0.70	0.79	0.80	
45		0.77	0.30	0.33	0.37	0.78	0.60	0.40	0.50	0.26	0.30	0.42	0.45	
40	0.58	0.80	0.68	0.45	0.34	0.36	0.42	0.89	0.66	0.72	0.60			
					Summer									
60° N		0.57	0.62	0.64	0.62	0.56	0.50	0.62	0.65	0.63	0.69	0.73	0.76	
55		0.43	0.56	0.52	0.45	0.51	0.57	0.60	0.52	0.55	0.70	0.76	0.78	0.80
52				0.58	0.48	0.44	0.47	0.56	0.64	0.66	0.67	0.70	0.70	0.70
45			0.72	0.59	0.50	0.58	0.65	0.65	0.63	0.61				
40	0.68	0.61	0.55	0.50	0.48	0.51	0.57	0.65	0.73	0.72	0.69	0.67	0.66	

spacing of 200 km, from correlation matrices in graphical form. Tables 5 and 6 list the variation in correlation coefficients at 60, 55, 45 and 40°N when $\rho = 200$ km and $\rho = 400$ km for summer and winter. It can be seen from Table 5 that, on the average, distance $\rho = 200$ km corresponds to $\mu = 0.70$. Here, for 60, 55 and 50°N the correlation coefficient fluctuates about the average level; for southern latitudes these fluctuations are somewhat larger. The lowest values of these coefficients are observed at 40°N $(\mu = 0.50$ when $\rho = 200$ km). The maximum values of these coefficients are observed at 60°N. The higher correlation in northern regions is apparently due to the predominance of large-scale circulation processes over local circulations in both winter and summer. In southern latitudes, intensive heating of individual regions results in the formation of local synoptic formations, which are hardly related to one another. It is noteworthy that the summer correlation function is smoother than the winter function at all latitudes. The same conclusion is confirmed by data of Table 6, which shows the variation in correlation functions by latitudes at $\rho = 400$ km. We note that the larger ρ, the greater the fluctuation of the correlation coefficient about the mean. When $\rho = 400$ km, the average value of μ is 0.6. In northern and central regions the variation in this correlation is smoother than in southern regions. The extremal values are $\mu = 0.99$ (40°N, 10°E) and $\mu = 0.20$ (52°N, 46°E).

Analysis of the correlation coefficients showed that substantial departures from homogeneity and isotropicity may also occur in the field of centered and normalized temperatures. As a rule, these departures occur in regions with a clearly expressed inhomogeneity of the underlying surface. Thus, data obtained at stations located on island or sea shores are hardly related to those of continental stations. Table 7 presents a temperature correlation matrix for weather ships and coastal stations. It is seen that the correlation between air temperatures in marine regions damps out much more rapidly than on land ($\sim$ 1000 km); this shows that the microclimatic disturbances are commensurable with synoptic disturbances.

To eliminate, at least partially, the inhomogeneity and anisotropicity, it is best to calculate correlation functions in an objective analysis by individual climatic regions, using some already available zoning system /1/. This can be found quite useful, particularly for the winter. We consider correlation functions for four regions: 1) central European USSR; 2) Northern USSR; 3) Western Europe (excluding southern regions); 4) southern regions of Western Europe.

According to /1/, these regions are close to climatic region zones. In calculating correlation functions the coefficients were averaged for all station pairs, the distance between which lies within the given graduation. The latter were selected in steps of 200 km. The correlation functions obtained for all four regions are listed in Tables 8 and 9. These tables also contain the number n of correlation coefficients averaged to obtain the corresponding correlation function. For comparison, Table 10 lists the correlation function for the whole of Europe.

TABLE 7. Temperature correlation matrix according to weather ships and coastal stations (summer)

	Coordinates		1	2	3	4	5	6	7
1	58° N	6° W	1.00 0	0.32 487	0.25 723	−0.10 1529	0.03 758	0.04 1024	0.17 929
2	54	10		1.00 0	0.28 248	0.13 1576	0.00 754	0.26 673	0.14 1400
3	52	10			1.00 0	0.26 1738	0.11 940	0.00 670	0.35 1648
4	62	33				1.00 0	0.22 831	0.05 1240	0.24 1687
5	59	19					1.00 0	0.26 650	0.11 1269
6	66	20						1.00 0	0.05 1820
7	66	1° E							1.00 0

N o t e. The first row gives the correlation coefficient, while the second lists the distance between stations (in kilometers).

TABLE 8. Correlation function by regions (summer)

Distance (km)	Region					
	1		2		3	
	μ	n	μ	n	μ	n
150	0.72	4	0.66	29	0.63	5
300	0.63	41	0.55	29	0.55	21
500	0.53	50	0.45	47	0.45	13
700	0.44	50	0.35	33	0.36	15
900	0.35	49	0.31	33	0.34	19
1100	0.26	43	0.23	24	0.28	13
1300	0.22	25	0.18	20	0.18	27
1500	0.14	28	0.10	15	0.07	8
1700	0.11	23	0.02	7	0.09	11
1900	0.10	17	0.00	4	0.08	4

TABLE 9. Correlation function by regions (winter)

Distance (km)	Region							
	1		2		3		4	
	μ	n	μ	n	μ	n	μ	n
150	0.76	30	0.71	13	0.74	22	0.59	12
300	0.73	117	0.66	70	0.63	122	0.45	53
500	0.61	132	0.56	100	0.56	149	0.42	57
700	0.53	142	0.44	100	0.51	181	0.49	44
900	0.47	129	0.36	100	0.41	185	0.44	44
1100	0.40	76	0.32	96	0.35	168	0.40	39
1300	0.41	65	0.27	78	0.36	154	0.34	40
1500	0.41	42	0.26	59	0.30	120	0.35	31
1700	0.44	23	0.18	53	0.27	91	0.22	32
1900	0.40	9	0.12	34	0.21	42	0.12	23

TABLE 10. Temperature correlation function for Europe (summer)

Distance (km)	μ	n	Distance (km)	μ	n
100	0.87	3 759	1700	0.19	78 293
300	0.65	24 351	1900	0.14	79 715
500	0.58	30 209	2100	0.08	80 147
700	0.50	47 106	2300	0.02	78 763
900	0.43	52 020	2500	—0.04	76 843
1100	0.35	67 638	2700	—0.08	75 232
1300	0.31	66 291	2900	—0.09	73 450
1500	0.25	76 993			

Note. The correlation function was calculated on the assumption that the field of centered and normalized temperatures is homogeneous and isotropic.

Calculations of correlation functions by regions showed that it is possible to assume homogeneity and isotropicity for the field of centered and normalized temperatures for three regions (1, 2 and 3). The somewhat unusual form of the correlation function for region 4 is apparently due to nonuniformity of the underlying surface and nonuniform distribution of the stations.

It can thus be concluded on the basis of an analysis of air-temperature correlations for Europe that the temperature field is generally inhomogeneous and anisotropic. Homogeneity and isotropy can be postulated for the field of centered and normalized temperatures in individual climatic regions. In the region with substantial underlying-surface nonuniformity even the field of centered and normalized temperatures is inhomogeneous and anisotropic.

Consider now some results of computing eigenvectors and eigenvalues of the temperature correlation matrices near the ground. In the example at hand we considered several groups of stations

located in different geographic regions of Europe.   The number of
stations was selected in such a manner as to provide for sufficient
spatial resolution for detecting synoptic and mesoscale
atmospheric disturbances /8/.   Each such group of stations can
correspond to some characteristic scale L of the temperature field
near the ground.   Here the number of stations is equal to the order
of the correlation matrices under study (Table 11).

TABLE 11

Area	Scale	Number of stations	Time of year
Europe	$L_1 \approx 10^3$ km	32	Winter, summer
European USSR	$L_2 \approx 10^2$ km	24	Summer
Central Europe	$L_2 < L_3 < L_1$	31	Winter
Northern Europe	$L_2 < L_3 < L_1$	29	"
Western Europe	$L_2 < L_3 < L_1$	32	"
Southern Europe	$L_2 < L_3 < L_1$	23	"

We calculated the eigenvectors of the matrices thus obtained.
Tables 12—14 list the first ten eigenvalues λ and residual variances
V_k, equal to $(\sum_{i=1}^{k} \lambda_i / \sum_{i=1}^{n} \lambda_i) \cdot 100$, for each matrix.   It is seen from common
analysis of the residual variances that the elimination of the small-scale
part of the spectrum in the starting information is highly effective —
the first ten eigenvectors account for, on the average, more than
80% of the total variance.   Screening-out of the small-scale part
of the spectrum (the greater the residual variance described by the
first vectors, the higher the screening effectiveness) is minimal in
summer (Table 12a, b, g, h).   This is particularly important when
studying the "reaction" of fields to atmospheric disturbances of a
given scale in different seasons /7/.   An investigation of the intermediate scales L_3 showed that when discussing a network of stations
with a resolution close to scale L_2, the effectiveness of screening-
out small-scale parts of the spectrum decreases rapidly for winter
too.

The stability of computing eigenvectors was studied using
various samples (50 in sampling 1 and 100 in sampling 2).   The
location and number of stations was the same $(n = 24)$ for each
sampling.   Table 12g was obtained from sampling 2 and Table 12i
from sampling 1.

It is seen from Table 13 that the sample size has a substantial
effect on the behavior of the vectors.   The fact that sampling 1
involved half the number of samples involved in sampling 2 points to
the fact that the second and all subsequent vectors differ markedly.

TABLE 12

k	λ	V_k	k	λ	V_k
a) Europe, winter $n=32$			e) Central Europe, winter $n=31$		
1	7.9	25.0			
2	4.0	37.5	1	15.8	51.0
3	3.3	47.8	2	4.2	64.5
4	2.4	55.1	3	2.5	72.6
5	2.0	61.4	4	1.6	77.8
6	1.8	67.4	5	1.5	82.7
7	1.5	72.1	6	1.1	86.3
8	1.3	76.2	7	0.85	89.0
9	1.2	80.0	8	0.83	91.6
10	1.1	83.6	9	0 71	93.9
			10	0.58	95.8
b) Europe, summer $n=32$					
			f) Western Europe, winter $n=32$		
1	6.5	20.5			
2	3.7	32.2	1	14.6	45.6
3	2.3	39.5	2	4.7	60.3
4	1.7	44.7	3	2.9	69.8
5	1.5	49.5	4	2.1	76.4
6	1.4	54.0	5	1.5	81.1
7	1.3	58.2	6	1.4	85.5
8	1.2	61.9	7	1.1	88.9
9	1.0	65.2	8	0.94	91.8
10	0.9	68.1	9	0 80	94.3
			10	0.61	96.2
c) European USSR, summer $n=24$			g) Western Europe, winter, $n=24$		
1	9.8	41.1	1	14.9	62.2
2	3 3	54.0	2	2.8	73.8
3	2.5	64.0	3	1.7	80.9
4	1.3	69.8	4	1.4	86.8
5	0.98	73.9	5	1.1	90.5
6	0.86	77.0	6	0.87	94.0
7	0 68	79.8	7	0.62	96.7
8	0.58	82.2	8	0.59	99.1
9	0.50	84.3	9	0.41	
10	0.40	86.0	10	0.35	
d) Southern Europe, winter $n=23$			h) Western Europe, summer, $n=24$		
1	9.5	41.1	1	9.8	40.6
2	3.6	56.7	2	3.4	54.9
3	1.9	65.0	3	1.8	62.5
4	1.6	72.0	4	1.5	68.9
5	1.5	78.5	5	1.0	73.2
6	1.2	83.7	6	0.89	76.9
7	1.1	88.5	7	0.83	80 4
8	0.76	91.8	8	0.74	83.5
9	0.62	93.5	9	0.67	86.3
10	0.51	95.7	10	0.50	88.4

TABLE 12 (continued)

k	λ	V_k	k	λ	V_k
i) Western Europe, winter $n = 24$			j) Northern Europe, winter $n = 29$		
1	8.4	35.1	1	12.0	41.5
2	5.8	59.1	2	3.6	53.9
3	2.0	67.7	3	3.1	64.6
4	1.5	73.9	4	2.3	72.5
5	1.0	78.0	5	1.4	77.3
6	0.82	81.4	6	1.2	81.4
7	0.62	84.0	7	1.0	84.8
8	0.57	86.3	8	0.94	88.0
9	0.53	88.6	9	0.90	91.0
10	0.51	90.7	10	0.77	93.8

TABLE 13

Station number	Vector					
	first		second		third	
	1	2	1	2	1	2
1	—0.29	—0.21	—0.06	—0.006	—0.18	—0.23
2	—0.24	—0.22	0.12	—0.06	—0.28	—0.14
3	—0.30	—0.27	0.05	—0.05	—0.27	—0.05
4	—0.19	—0.20	0.24	—0.12	—0.07	—0.06
5	—0.15	—0.16	0.23	—0.26	0.18	—0.21
6	—0.15	—0.22	0.29	—0.20	0.30	—0.16
7	0.004	—0.21	0.35	—0.31	0.16	—0.25
8	0.003	—0.19	0.32	—0.08	0.24	0.27
9	—0.04	—0.18	0.30	—0.12	—0.17	—0.005
10	—0.07	—0.26	0.30	—0.11	0.23	0.08
11	—0.12	—0.24	0.23	—0.15	0.19	0.06
12	—0.18	—0.28	0.22	—0.05	0.14	0.06
13	—0.21	—0.22	0.23	—0.09	—0.13	0.24
14	—0.29	—0.25	0.03	0.12	—0.17	0.08
15	—0.29	—0.20	—0.02	0.11	0.02	0.05
16	—0.31	—0.24	—0.06	0.18	—0.05	0.14
17	—0.27	—0.22	—0.09	0.24	—0.004	0.18
18	—0.23	—0.06	—0.13	0.41	0.09	0.16
19	—0.27	—0.17	—0.15	0.24	0.06	0.17
20	—0.21	—0.14	—0.09	0.34	0.04	—0.04
21	—0.17	—0.22	—0.19	0.15	0.23	—0.01
22	—0.23	—0.19	—0.14	0.13	0.21	—0.004
23	—0.13	—0.02	—0.23	0.44	0.43	—0.45
24	—0.09	—0.09	—0.21	0.16	0.34	—0.57

It is also worthwhile considering the stability of expansions relative to the number of stations. For this we selected (to scale L_1) sections along latitude and longitude lines, as well as along intermediate directions. The correlation matrix was constructed for stations located along such a section and eigenvectors were calculated. Then one station was eliminated from the original matrix and the eigenvectors were recalculated. Table 14 lists the values of the components of the first eigenvector for a section along 52°N (the "edge" stations 1 and 11 have coordinates 52°N, 8°W

TABLE 14

Number of eliminated stations	Station number										
	1	2	3	4	5	6	7	8	9	10	11
—	0.16	0.38	0.43	0.39	0.36	0.13	—0.27	—0.12	—0.32	—0.35	—0.16
1	—	0.36	0.42	0.38	0.37	0.12	—0.28	—0.14	—0.34	—0.36	—0.16
2	0.12	—	0.35	0.37	0.35	—0.06	—0.36	—0.28	—0.41	—0.44	—0.23
3	0.13	0.30	—	0.35	0.33	0.00	—0.40	—0.24	—0.43	—0.46	—0.26
4	0.14	0.35	0.40	—	0.32	0.03	—0.36	—0.20	—0.32	—0.34	—0.23
6	0.15	0.36	0.42	0.38	0.36	—	—0.30	—0.15	—0.34	—0.37	—0.18
7	0.18	0.40	0.47	0.43	0.40	0.22	—	—0.05	—0.26	—0.30	—0.20
8	0.16	0.38	0.45	0.41	0.38	0.16	—0.25	—	—0.28	—0.33	—0.15
9	0.21	0.43	0.48	0.44	0.41	0.21	—0.20	—0.05	—	—0.28	—0.11
10	0.22	0.45	0.49	0.45	0.42	0.22	—0.19	—0.06	—0.24	—	—0.10
11	0.17	0.31	0.45	0.41	0.38	0.15	—0.25	—0.10	—0.30	—0.33	—
2, 4, 5, 6, 8, 9	0.26	—	0.43	—	—	—	—0.40	—	—	—0.59	—0.42

and 52°N, 60°E, respectively). Elimination of one or even several stations does not result in a significant redistribution of the values of the components of the first vectors. It should be noted that minima are observed when the "edge" stations are eliminated. A similar result was obtained at all the remaining sections, so confirming claims that the expansions with respect to natural orthogonal components /7/ are stable. Calculations showed that the eigenvectors of correlation matrices composed along sections of the field are in sufficient agreement with expansions of two-dimensional fields, provided the resolution of the field equals that of the section.

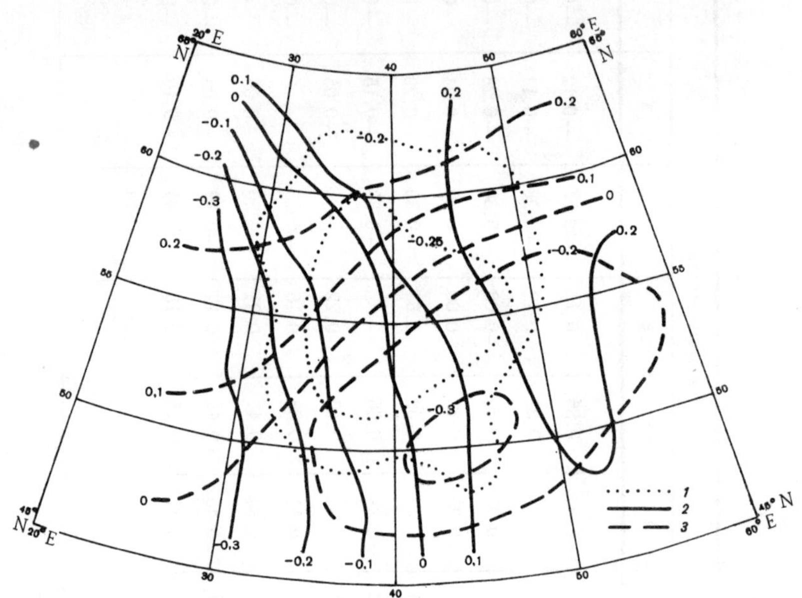

FIGURE 1. First three eigenvectors of the temperature correlation matrix for European USSR:

1, 2 and 3 — first, second and third vectors.

One of the indications of the homogeneity of a field of some random variable is nearness of the eigenvectors of the correlation matrix to trigonometric functions /5/. Figure 1 represents the first three eigenvectors of the temperature correlation matrix for European USSR. The first vector is virtually constant, while the second and third vectors fluctuate sinusoidally. This result can be regarded as some confirmation of the homogeneity and isotropicity of the underlying-surface temperature field in the corresponding scale.

These results can be compared with those cited in /6, 7, 12, 13/.
A quantitative comparison was carried out only for scale L_1 with
data of /6, 7/. Grimmer /12/ resolved monthly surface temperature anomaly data of 32 stations over 80 years. In spite of the fact
that the part of Europe considered by Grimmer is located somewhat
to the west of that considered here, the isolines of the "overlapping"
regions are quite similar to one another. This is clearly a
qualitative conclusion, since no numerical characteristics of the
eigenvector isolines are presented in /12/.

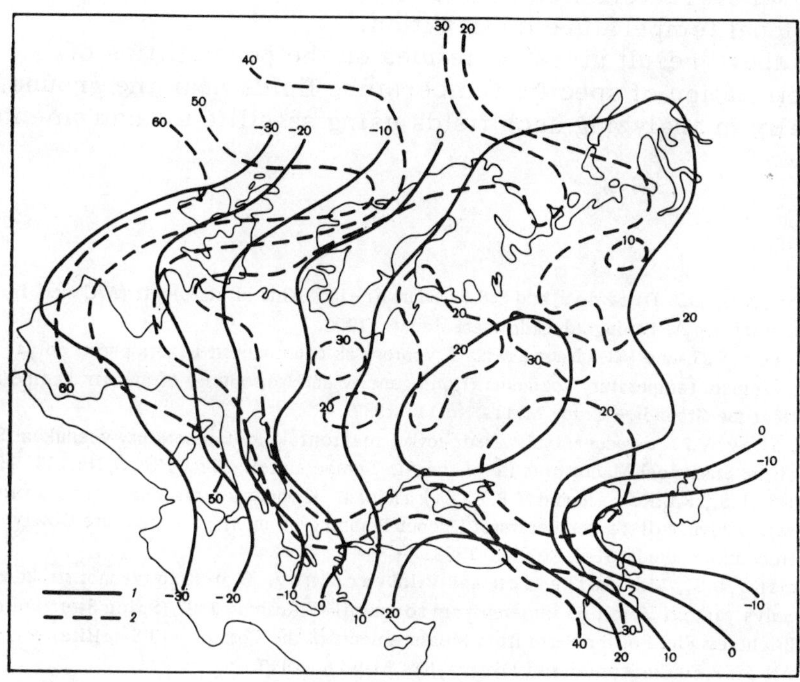

FIGURE 2. First eigenvector of the temperature correlation vector for Europe (1)
and probability (%) of the deviation of the monthly average temperature by ±1°
from the normal (2) /6/.

Of the 26 stations of the first sector of the northern hemisphere
treated in /7/, 13 stations are contained within the limits of scale
L_1 considered by us. Consequently, L_1 can be regarded as a
"denser" network of stations "inscribed" in the 1st sector.
Comparison shows that both resolutions are virtually identical.
The natural orthogonal components describing large-scale features
of the temperature field near the ground over a large area were
thus shown to react little to any change in the number and location

of stations, and also to changes in the time over which the initial data are averaged. (Fields of daily and monthly average temperature data of 26 stations for 58 years were considered in /7/.)

Natural orthogonal components correspond somewhat to specific features. Figure 2 shows isolines of the first eigenvector and the probability of departure of the monthly average air temperature by ± 1°C from the normal according to data of /6/. The second, third and fourth vectors similarly correspond to the probability that the monthly average temperature deviates by ± 2, 3 and 4°C from normal, respectively. This correspondence may not be random, since both characteristics are in effect the parameters of the multi-dimensional temperature distribution.

The above result gives some idea on the possibilities of parametrization of specific temperature fields near the ground, especially in analyzing such fields using satellite measurements.

BIBLIOGRAPHY

1. Alisov,B.P., O.A.Drozdov, and E.S.Rubinshtein. Kurs klimatologii (A Course in Climatology). Leningrad, Gidrometeoizdat. 1952.

2. Boldyrev,V.G. and V.I.Khamarin. K voprosu ob odnorodnosti i izotropnosti polya prizemnoi temperatury vozdukha (Homogeneity and Isotropicity of the Air Temperature Near the Ground).— Trudy MMTs, No.11. 1967.

3. Boltenkov,V.P. Issledovaniya statisticheskoi makrostruktury temperatury vozdukha (Studies of the Statistical Macrostructure of the Air Temperature).— Trudy GGO, No.165. 1965.

4. Gandin,L.S., R.L.Kagan, and V.P.Tarakanova. K voprosu o ratsional'nom planirovanii seti nablyudenii za temperaturoi (Proper Planning of an Air-Temperature Observation Network).— Trudy GGO, No.228. 1968.

5. Istomina,L.G., M.S.Malkevich, and V.I.Syachinov. O prostranstvennoi strukture polya yarkosti Zemli po izmereniyam so sputnika "Kosmos-149" (Space Structure of the Brightness Field of the Earth from Measurements of the Cosmos-149 Satellite).— Izvestiya AN SSSR, Fizika Atmosfery i Okeana, Nos.5 and 6. 1970.

6. Lebedev,A.N. Veroyatnosti otkloneniya srednei mesyachnoi temperatury vozdukha ot mnogoletnei srednei po territorii Evropy i SSSR (Atlas kart) (Probability of Deviations in the Monthly Average Air Temperature for Europe and the USSR (Atlas of Maps)). Leningrad, Gidrometeoizdat. 1957.

7. Meshcherskaya,A.V., et al. Estestvennye ortogonal'nye sostavlyayushchie meteoro-logicheskikh polei (The Natural Orthogonal Components of Meteorological Fields). Leningrad, Gidrometeoizdat. 1970.

8. Monin,A.S and A.M.Yaglom. Statisticheskaya gidromekhanika (Statistical Fluid Mechanics), Part 2.— Moskva, Izdatel'stvo "Nauka." 1965.

9. Myach,L.T. O statisticheskikh kharakteristikakh i ob"ektivnom analize polya vlazhnosti i temperatury u poverkhnosti Zemli (Statistical Characteristics and Objective Analysis of the Humidity and Temperature Field at the Ground Surface).— Trudy MMTs, No.7. 1965.

10. Khatamkulov,G. Ob ob"ektivnom analize polya otnositel'nogo potentsiala nad okeanami (Objective Analysis of the Relative Potential Field Above Oceans).— In: Sbornik "Dinamicheskaya meteorologiya." Tashkent. 1965.

11. H r d a , J. O statisticheskoi strukture prizemnogo polya temperatury vozdukha na territorii
        Chekhii i Moravii (Statistical Structure of Atmospheric Temperature Field near the
        Ground in Bohemia and Moravia).— Idöjárás, Budapest, Vol.72, pp.210—215. 1968.
12. G r i m m e r , M. The Space-Filtering of Monthly Surface Temperature Anomaly Data in Terms
        of Pattern, using Empirical Orthogonal Functions.— Quart. J. Roy. Met., Soc., Vol.89,
        pp.395—408. 1963.
13. K u t z b a c h , J.E. Empirical Eigenvectors on Sea-Level Pressure, Surface Temperature and
        Precipitation Complexes over North America.— J. Appl. Meteor., Vol.6, pp.791—802.
        1967.

UDC 551.509:551.521.32

EFFECTIVENESS OF UTILIZING CLOUDINESS DATA OBTAINED FROM SATELLITES IN OBJECTIVE ANALYSIS OF THE WIND FIELD

L. A. Anekeeva

A program is presented for a numerical analysis of the wind field using satellite cloud information, and it is shown by means of several examples that the utilization of such information in numerical analysis of the wind field gives satisfactory results.

The technique of reconstructing the wind field from satellite photographs of the cloud cover in the region of a cloud vortex, and also preliminary results on the utilization of this technique in calculating U and V at nodes of a rectangular grid have been treated elsewhere /3, 4/. This paper describes a number of calculations for clarifying whether utilization of satellite information results in improving the analysis of the wind field in regions with a sparse network of aerological stations.

We carried out the following experiment. Several computations were conducted using a program for objective analysis of the wind field involving polynomial interpolation /6/. First, only wind data collected by aerological stations were used. Then, in regions where satellite photographs show a cloud vortex, all these values of the wind direction and speed were replaced by values obtained in line with the technique of /3, 4/. Then the calculations were carried out in such a manner that no wind data were available in the region of the vortex. In the latter case the program utilizing polynomial interpolation did not function, and hence it was necessary to use climatic data on the wind taken from an atlas /5/.

Calculations were made only when a clearly defined cloud vortex occupying a sufficiently wide area was located over a region with a sufficient aerological-data collection network. Twelve such cases were selected, including vortices photographed from Meteor and ESSA satellites. The results were compared for those nodes of the rectangular grid which covered the cloud-vortex region. There was a total of 980 such points.

Results of the analysis carried out with satellite and climatic wind data were compared with results obtained by analyzing data of the radiosonde network. We found the mean absolute errors (a_s, a_{cl}), the rms deviations (σ_s, σ_{cl}), as well as values of quantity ρ (ρ_s, ρ_{cl}), equal to the ratio of the number of points at which the calculated and actual U and V had the same sign to the total number of points (subscript "s" pertains to quantities obtained on the basis of satellite data, while subscript "cl" is the same for climatic data). The average estimates obtained are tabulated in Table 1, from which it is clear that the results of objective analysis carried out from satellite data are in all respects better than those obtained from the results of analyzing climatic wind data.

TABLE 1. Estimates of the results of computing U and V by a program utilizing polynomial interpolation of starting data, $n = 980$

P mb	U						V					
	a_s	a_{cl}	σ_s	σ_{cl}	ρ_s	ρ_{cl}	a_s	a_{cl}	σ_s	σ_{cl}	ρ_s	ρ_{cl}
1000	1.46	2.09	2.00	2.84	0.89	0.78	1.55	1.78	2.10	2.42	0.87	0.82
850	2.11	3.39	2.75	4.41	0.89	0.83	2.44	2.56	3.22	3.43	0.85	0.80
700	2.47	3.89	3.23	5.07	0.88	0.80	2.40	3.22	3.14	4.22	0.86	0.81
500	2.69	4.69	3.46	6.14	0.93	0.80	2.54	3.60	3.48	4.97	0.88	0.83
300	3.90	5.91	5.07	7.59	0.90	0.84	3.13	4.19	4.25	5.78	0.87	0.82

Then, similar calculations for the same time periods were also carried out using a program for objective analysis of the wind field, constructed on the basis of principles of optimal interpolation /11/.

TABLE 2. Estimates of the results of computing U and V by a program of objective analysis utilizing principles of optimal interpolation of starting data, $n = 980$

P mb	U						V					
	a_s	$a_{n.d}$	σ_s	$\sigma_{n.d}$	ρ_s	$\rho_{n.d}$	a_s	$a_{n.d}$	σ_s	$\sigma_{n.d}$	ρ_s	$\rho_{n.d}$
850	2.23	3.00	2.65	5.00	0.94	0.87	2.41	2.91	4.07	4.51	0.94	0.89
500	3.31	4.26	4.36	6.77	0.94	0.88	3.30	4.64	4.72	5.40	0.92	0.87
300	4.02	6.43	6.31	8.64	0.92	0.85	4.11	5.93	5.45	6.73	0.91	0.85

Table 2 lists the mean absolute errors, rms deviations and average values of ρ, obtained from satellite data and with no available wind data (subscripted n.d). Table 2 shows that the results of objective analysis carried out from satellite data are better than those obtained without wind data.

Comparison of Tables 1 and 2 indicates that the corresponding quantities are close to one another. Hence in subsequent calculations we used only the program with polynomial interpolation and conducted calculations for a large number of isobaric surfaces. When the comparison required calculating values of U and V at grid nodes with no wind data over a large area, we used the climatic values of U and V.

Relationships given in /3, 4/ for relating air-flow and cloudiness fields in a cyclone, as well as known relationships between the wind and cloudiness fields in the presence of other cloud structures, photographed by satellites /1, 8—10, 12/, were utilized in developing the scheme of objective analysis of the wind field. This was based on the scheme of objective analysis of the wind field developed in /6/. The program consists of two parts: an auxiliary part, where the input information is deciphered and reduced to a form conveni- ent for interpolation; and a main part, where the values of U and V proper at stations are interpolated to nodes of the rectangular grid. The main program was retained almost unchanged, while the auxiliary program was completely revised.

In our case the input data are those from radiosonde observations of stationary aerological stations, as well as satellite data pertain- ing to arbitrary points with geographical coordinates φ and λ. The sequence in which stations and auxiliary points are introduced is arbitrary.

During operation of the program the aerological-station indexes are replaced by the coordinates of the given station in the rectangular grid, while the geographical coordinates of the auxiliary points φ, λ (subsequently these will also be termed stations) are converted by the formulas /2/.

$$x = a + 39.62762 \, \frac{1 - \operatorname{tg}\frac{\varphi}{2}}{1 + \operatorname{tg}\frac{\varphi}{2}} \sin\left(\lambda - \frac{\pi}{4}\right),$$

$$y = b - 39.62762 \, \frac{1 - \operatorname{tg}\frac{\varphi}{2}}{1 + \operatorname{tg}\frac{\varphi}{2}} \cos\left(\lambda - \frac{\pi}{4}\right)$$

to rectangular coordinates (x, y). Here a and b are the coordinates of the pole in the rectangular system.

All the starting data are then arranged in some order: the entire region is subdivided into strips parallel to the X axis (the width of the strips is equal to the grid spacing), and within each strip the stations are numbered in the order of increasing abscissas.

The following stage involves deciphering the input information. When radiosonde network data are available on the direction and speed of the wind at five isobaric surfaces, which occupy at input three memory cells, they are decoded in such a manner that the wind direction is expressed in degrees and the velocity in m/sec.

When satellite data are available, all the necessary input data occupy a single cell, which gives the structure of the cloudiness and direction of the cloud band, and in the case of a cloud vortex the number of the sector, distance between station and vortex center, and the latter's radius. From these data, using tables both of the angle $\Delta\varphi$ between the cloud-band direction and the wind direction and of the wind speed f for different cloud structures at five isobaric surfaces, one determines $\Delta\varphi$ and f for each station; the wind direction at the given station and at the given level is then determined from the expression $d = f_{band} + \Delta\varphi$.

Now the wind directions and speeds are available for each station at each of the five isobaric surfaces. Hence one calculates the horizontal components of the wind vector along an orthogonal reference system:

$$U = f\cos\psi; \quad V = f\sin\psi',$$

where f is the wind speed (m/sec); $\psi = \gamma - d$, d being the wind direction (deg); $\gamma = 270 + \delta$ (δ is the angle through which the ordinate axis deviates from the meridian for the given station and is a function of the latter's longitude and is determined during computations).

The auxiliary program is now used to process the starting data to a form convenient in the interpolation of values of U and V at stations to nodes of the rectangular grid. This interpolation is carried out utilizing the main program, which in this work remains almost unchanged and is therefore not described. The accuracy of data obtained by this program lies within tenths of m/sec.

A number of experiments were conducted to prove the utility of satellite data in analyzing the wind field over areas with sparse aerological networks. The program of /2/ was used to calculate two-dimensional streamlines from values of U and V, obtained as a result of analysis using three versions: 1) the wind field analysis was carried out only with data of the radiosonde network; 2) it was assumed that no wind data exist in regions with some characteristic cloudiness structure, and so climatic values of U and V were used; 3) only satellite data were used for regions with some

characteristic cloudiness structure. Calculations using the first
two versions were carried out with the program described in /6/,
while those utilizing the latter version used the program described
above.

The streamlines thus calculated were drawn directly on the
cloud picture. In regions with a dense radiosonde network
satisfactory agreement was found between streamline fields constructed from satellite and aerological data. In all the calculated
examples there was no case when streamlines constructed from
radiosonde data pointed to convergence of flow toward the vortex
center, while those constructed from satellite data indicated
divergence of flow. On the other hand, although it is sufficient that
a streamline deviates only slightly from the real direction in order
to obtain an incorrect pattern, there is no similarity between
streamline fields constructed from climatic values of the wind
components and the real flow pattern.

To illustrate the above we present the following example.
Figure 1 shows a mosaic obtained by the ESSA-6 satellite on
27 September 1968. On it are drawn streamlines calculated
according to the first (Figure 1a) and third (Figure 1b) versions on
the 700-mb surface. This mosaic is of interest in that it shows two
cloud vortices, one of which is situated over European USSR for
which sufficient aerological data are available, while the other is
situated over the North Atlantic where such data are sparse. It is
seen that in the vortex region over European USSR agreement
between the pattern of streamlines calculated from satellite data and
and that obtained from radiosonde observations is satisfactory.
This is, however, not true of the North Atlantic. The point is that
very little wind data were available at the given time for the North
Atlantic (Figure 1a contains all available data on wind direction and
speed for the region under study), and these did not indicate any
eddying flow. Calculations from satellite data, on the other hand,
point to the existence in this region of rotating air motion, which
is confirmed by pressure-surface maps (Figure 2a).

This example also confirms that utilization of satellite data in
analyzing the wind field over regions with a sparse network of
aerological stations yields better results than the use of an
insufficient amount of real wind data. On the other hand, in regions
with a dense station network satellite data yield results close to
the actual state of affairs.

Still another experiment which confirms the above conclusion is
the reconstruction of the fields of geopotential H from wind data.
The technique used was that of /7/. Calculations by means of the
M-20 computer for a 24 X 20 network with 300-km spacing showed

that the errors in reconstructing the field of *H* at the 700-mb level
from satellite wind data for regions with a dense station network
are somewhat higher than those obtained using the actual wind. On
the other hand, in regions with a sparse network, reconstruction
from satellite data yields much better results than when using the
actual wind.

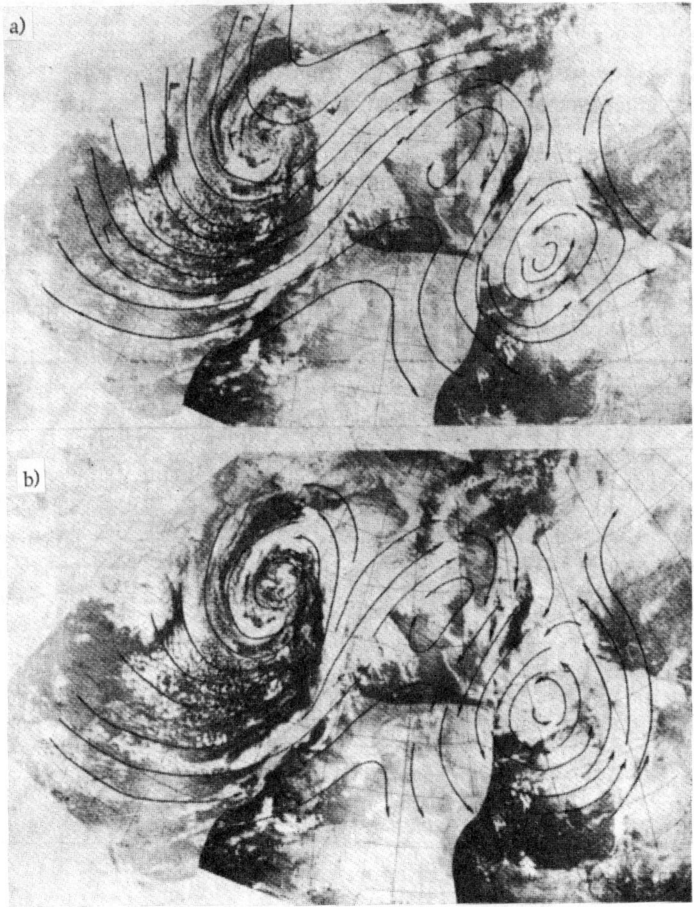

FIGURE 1. Streamlines at the 700-mb surface on 27 September 1968
calculated from radiosonde (a) and satellite (b) data.

Figures 2a and b are presented as examples. Figure 2a shows
the 700-mb chart at 1500 hrs Moscow time on 27 September 1968.
The chart clearly shows two cyclones, one centered in the Krivoi
Rog region and the other at a point with coordinates 56°N, 13°W.

These cyclones correspond to the two cloud vortices shown in
Figure 1. Figure 2b depicts the field of *H* at the 700-mb isobaric
surface reconstructed from values of *U* and *V*, obtained as a result
of objective analysis of radiosonde data alone. It is seen that the
cyclone above European USSR has been reconstructed quite
satisfactorily, the average error in reconstructing *H* in the vicinity
of this cyclone (the averaging region is delineated by a dashed line)
being 1.71 decameters. At the same time there is no cyclone over
the Atlantic. A deep trough does exist, but there are no closed
isohypses. The error in reconstructing *H* in the region of this
cyclone (also delineated on the figure) was 4.21 decameters.

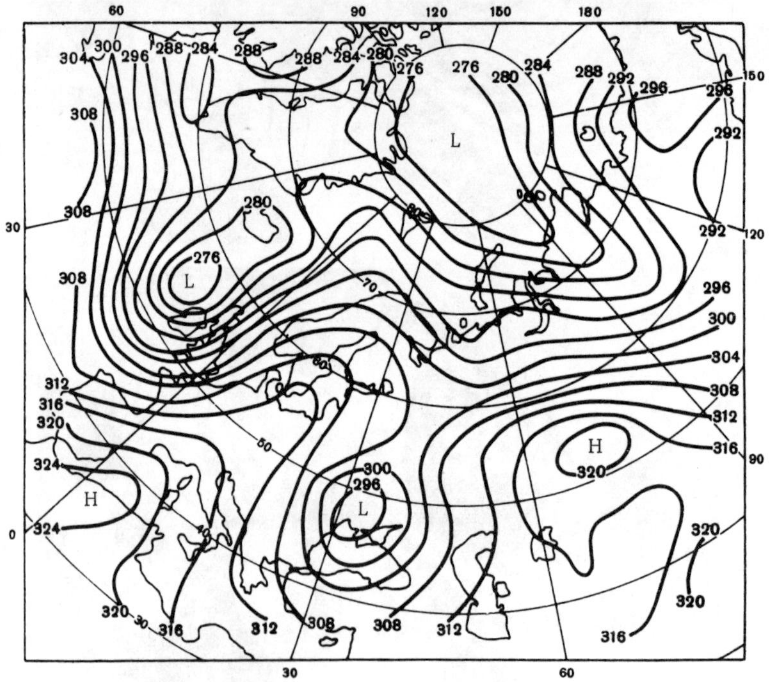

FIGURE 2a. Map of actual values of geopotential *H* at 1500 hrs Moscow time
on 27 September 1968.

Figure 2c displays the field of *H* reconstructed from values of
U and *V* obtained from objective analysis utilizing satellite data.
Here only cloud data were used in regions of cloud vortices. It is
seen that in the region of the cyclone above the European USSR,
where a dense network of aerological sections exists, the mean
error in reconstructing *H* is 2.54 decameters, which is somewhat
poorer than when using actual wind data. On the other hand, the

pattern above the Atlantic in the region of the cyclone was much
better.   The map shows a cyclone, the center of which coincides
approximately with the center of the real cyclone shown in
Figure 2a.   The mean error in reconstructing H was 2.84
decameters.

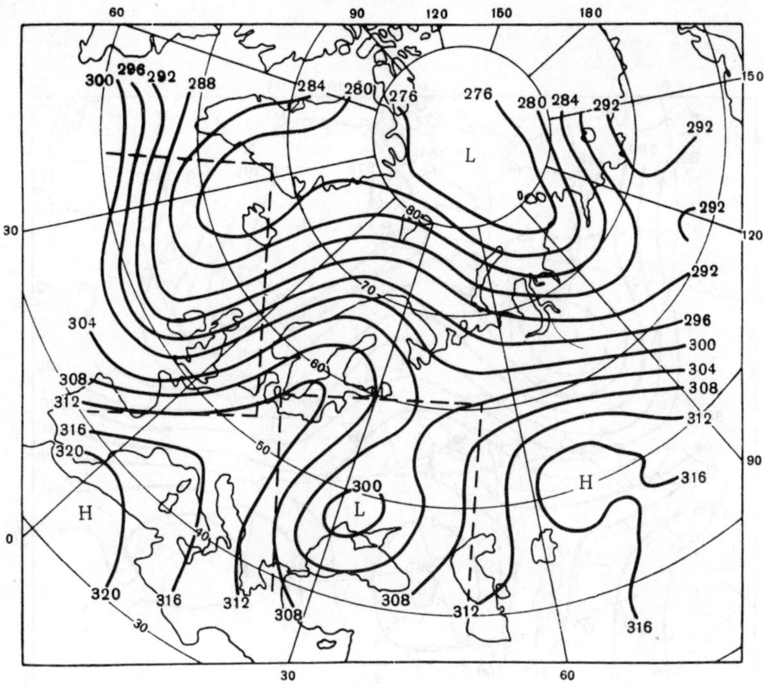

FIGURE 2b. Map of values of H reconstructed from the real wind at 1500 hrs
Moscow time on 27 September 1968.

Since the purpose of meteorological studies is to forecast
meteorological elements, an attempt was made to evaluate the
effectiveness of using satellite information in forecasting the wind
field.   We used in our calculations the scheme for forecasting the
wind and geopotential fields developed in /13/, which uses as input
data the values of the geopotential and wind components at the
700-mb surface at nodes of a 26 × 22 grid with a 300-km spacing.
The program also provides for the use of the geostrophic wind.
The forecast is for one day ahead.
Twelve daily forecasts were calculated.   Several versions were
calculated in each case: the first version used as starting data
wind information obtained as a result of objective analysis of the

radiosonde data alone; the second forecast was carried out as a
result of objective analysis utilizing, in regions of cloud vortices,
only satellite cloud data; the third was obtained as a result of
objective analysis using in regions of cloud vortices only climatic
data on the wind; in the fourth, geostrophic relationships of the
wind were used as starting wind data.   The values of H in all four
versions are identical.

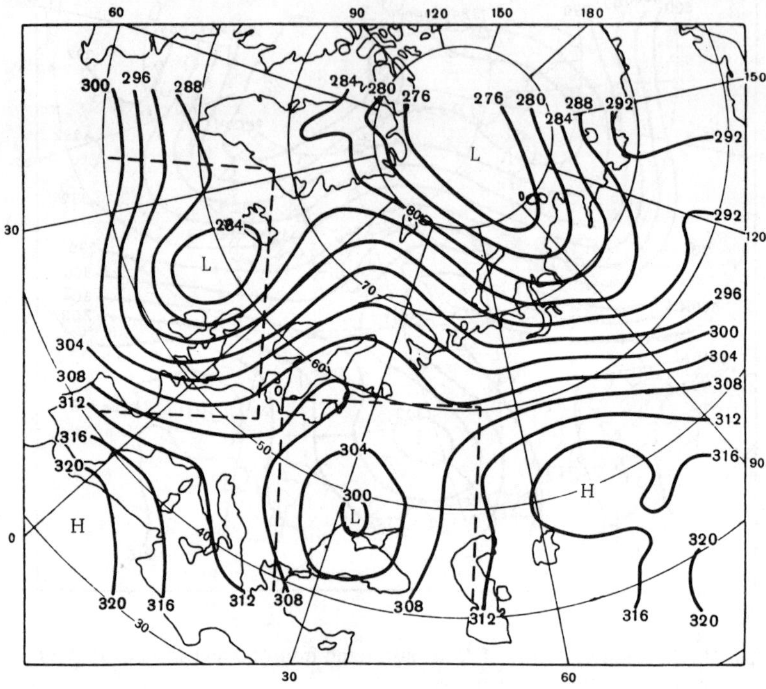

FIGURE 2c. Map of values of H reconstructed using satellite cloud data at
1500 hrs on 27 September 1968.

The forecasts were evaluated for an internal region of 14 × 10.
We calculated the mean relative forecasting errors ε for values of
H, U and V, equal to the ratio of the mean absolute error to the mean
absolute variation of this quantity.   We also calculated values of ρ,
which is an estimate of the agreement between the signs of the
fields of the actual and forecast variations, and is equal to the ratio
of the difference between the number of agreements in the signs of
variations and the number of their disagreements to the total
number of points being estimated.

The estimates are tabulated in Table 3, from which it is seen that the forecast of weather components made one day ahead using satellite data is better than forecasting based on the geostrophic wind and much better than that based on climatic data.

TABLE 3. Estimates of the quality of geopotential and wind-component fields at the 700-mb level using different starting data

Starting data	U		V		H	
	ε	ρ	ε	ρ	ε	ρ
Geostrophic wind	0.78	0.81	0.81	0.70	0.76	0.59
Radiosonde data	0.69	0.84	0.76	0.72	0.74	0.59
Satellite data	0.70	0.84	0.76	0.72	0.74	0.59
Climatic data	0.87	0.77	0.88	0.65	0.95	0.49

Hence the wind field reconstructed from satellite cloud photographs using the technique described here can be used as additional information in objective analysis of the wind field; all the results obtained are satisfactory.

BIBLIOGRAPHY

1. Alekseeva, I.A. and N.F.Vel'tishchev. Orientatsiya otnositel'no vozdushnogo potoka gryad oblachnosti, nablyudaemykh s iskusstvennykh sputnikov Zemli (Orientation of Cloud Masses Observed from Artificial Satellites Relative to the Air Flow).— Trudy MMTs, No.11. 1966.
2. Anekeeva, L.A. and I.P. Vetlov. Skhema rascheta i analiz prostranstvennykh linii toka i traektorii vozdushnykh chastits (Computational Scheme and Analysis of Three-Dimensional Streamlines and Air-Particle Trajectories).— Trudy MMTs, No.11. 1966.
3. Anekeeva, L.A. Ispol'zovanie dannykh ob oblachnosti, poluchennykh s pomoshch'yu ISZ, v ob"ektivnom analize polya vetra (Utilization of Satellite Cloud Data in Objective Analysis of the Wind Field).— Trudy Gidromettsentra SSSR, No.36. 1968.
4. Anekeeva, L.A. K voprosu ob ispol'zovanii sputnikovykh dannykh ob oblachnosti v ob"ektivnom analize polya vetra (Utilization of Satellite Cloud Data in Objective Analysis of the Wind Field).— Trudy Gidromettsentra SSSR, No.50. 1969.
5. Guterman, I.G. (editor). Aeroklimaticheskii atlas kharakteristik vetra severnogo polushariya (Aeroclimatic Atlas of Characteristics of Northern-Hemisphere Winds). NIIAK. 1963.
6. Bykov, V.V., G.P.Kurbatkin, and I.V.Gorelysheva. Opyt postroeniya mnogourovennoi skhemy chislennogo analiza aerologicheskikh dannykh (Experience in Constructing a Multilevel Scheme for Numerical Analysis of Aerological Data).— Trudy MMTs, No.4. 1964.
7. Burtsev, A.I. and I.P.Vetlov. Vosstanovlenie polya geopotentsiala po polyu vetra i polya vetra po vikhryu i divergentsii (Reconstruction of the Geopotential Field from the Wind

Field and of the Wind Field from the Curl and Divergence).— Meteorologiya i Gidrologiya, No.5. 1962.

8. Vel'tishchev, N.F. and L.I.Silaeva. Konvektivnye oblachnye yacheiki po nablyudeniyam s iskusstvennykh sputnikov Zemli (Convective Cloud Cells According to Satellite Observations).— Trudy Gidrometeorologicheskogo Tsentra SSSR, No.20. 1968.

9. Vel'tishchev, N.F. Yacheikovaya konvektsiya v atmosfere (Cellular Convection in the Atmosphere).— Trudy Gidrometeorologicheskogo Tsentra SSSR, No.50. 1969.

10. Dzyubenko, G.D. and A.M.Tsar'kova. Opredelenie polozheniya osi struinogo techeniya po dannym ob oblachnosti s meteorologicheskogo sputnika (Determination of the Location of the Jet-Stream Axis from Data of a Meteorological Satellite).— Trudy Gidrometeorologicheskogo Tsentra SSSR, No.36. 1968.

11. Krichak, M.P. Metodika ob"ektivnogo analiza polya vetra na urovnyakh 850, 500 and 300 mb (A Technique of Objective Analysis of the Wind Field at the 850, 500 and 300 mb Levels).— Meteorologiya i Gidrologiya, No.1. 1969.

12. Smirnova, N.V. Volnistye oblaka prepyatstvii, nablyudaemye s iskusstvennykh sputnikov Zemli (Wave-Shaped Clouds Produced by Obstacles and Observed from Artificial Satellites).— Trudy Gidrometeorologicheskogo Tsentra SSSR, No.20. 1968.

13. Tskvitinidze, Z.I. K chislennomu prognozu geopotentsiala i vetra (Numerical Forecasting of the Geopotential and Wind).— Meteorologiya i Gidrologiya, No.2. 1969.

UDC 551.513.2:551.501.776

CALCULATION OF GEOPOTENTIAL FIELDS AT VARIOUS ATMOSPHERIC LEVELS FROM DATA ON THE OVERALL CLOUDINESS AND TEMPERATURE

A. I. Burtsev, Sh. A. Musaelyan

A method is presented for calculating the geopotential fields at various atmospheric levels from the 500-mb field and from the temperature distribution. The 500-mb chart used was reconstructed from overall cloud cover data obtained from meteorological satellites. Computational examples are given.

Existing methods for reconstructing the geopotential field from the overall cloudiness can only be used at a single atmospheric level /1, 2/. This brief paper treats the utilization of these results as well as temperature data for reconstructing geopotential fields for other atmospheric levels. As is known, if the geopotential field at one atmospheric level and the temperature distribution everywhere are known, it is possible to calculate the geopotential field at any other level.

Suppose, for example, the geopotential field of the 500-mb surface and the temperature distribution at the 850, 700, 500 and 300 mb surfaces are known and denoted by H_{500}, T_{850}, T_{700}, T_{500} and T_{300}, respectively; functions describing the geopotential fields at the 850, 700, and 300 mb surfaces are denoted by H_{850}, H_{700} and H_{300} respectively. Elementary transformations of the polytropic formula yield relationships relating the unknown functions H_{850}, H_{700} and H_{300} to the known functions H_{500}, T_{850}, T_{700} and T_{300}:

$$H_{300} = H_{500} + \frac{2.9255\,T_{500}\ln\frac{5}{3}}{1 + \frac{\varkappa_3}{2} + \frac{\varkappa_3^2}{3} + \frac{\varkappa_3^3}{4} + \frac{\varkappa_3^4}{5}},$$

$$H_{700} = H_{500} + \frac{2.9255\,T_{500}\ln\frac{5}{7}}{1 + \frac{\varkappa_7}{2} + \frac{\varkappa_7^2}{3} + \frac{\varkappa_7^3}{4} - \frac{\varkappa_7^4}{5}}, \qquad (1)$$

$$H_{850} = H_{700} + \frac{2.9255\,T_{700}\,\ln\frac{7}{8.5}}{1 + \frac{\varkappa_{8.5}}{2} + \frac{\varkappa_{8.5}^2}{3} + \frac{\varkappa_{8.5}^3}{4} + \frac{\varkappa_{8.5}^4}{5}}, \qquad (1)$$

where

$$\varkappa_3 = 1 - \frac{T_{300}}{T_{500}},$$

$$\varkappa_7 = 1 - \frac{T_{700}}{T_{500}},$$

$$\varkappa_{8.5} = 1 - \frac{T_{850}}{T_{700}}. \qquad (2)$$

These formulas were used to calculate the fields of H_{300}, H_{700} and H_{850}, where H_{500} was calculated from TIROS-VII cloud data for 29 August 1963 /1/, and T_{300}, T_{500}, T_{700} and T_{850} were the corresponding temperature data at 1500 hrs for 29 August 1963. The recurrence of the absolute error and chart-average absolute error of the calculated fields is listed in Table 1.

TABLE 1. Recurrence of the absolute error and chart-average absolute error for calculated 300, 700 and 850 mb charts

Chart	Absolute error (decameters)									
	0	1	2	3	4	5	6	7	8	9
300 mb	3	5	4	7	6	12	10	16	19	16
700 mb	17	42	41	37	40	22	19	16	19	13
850 mb	20	48	39	33	31	30	22	16	10	13

Chart	10	11	12	13	14	15	16	17	18	Mean absolute error (decameters)
300 mb	23	24	21	19	25	25	20	16	49	12.5
700 mb	6	12	10	7	11	5	2	1	—	5.3
850 mb	12	9	14	11	5	4	2	1	—	5.2

The recurrence for calculated H_{700} and H_{850} is high for small errors and low for large errors. The chart-average absolute error is here 5.3 and 5.2 decameters, respectively. The recurrence of the absolute error for the field of H_{300} is high for large errors and low only for small errors; the chart-average error here is 12.5 decameters.

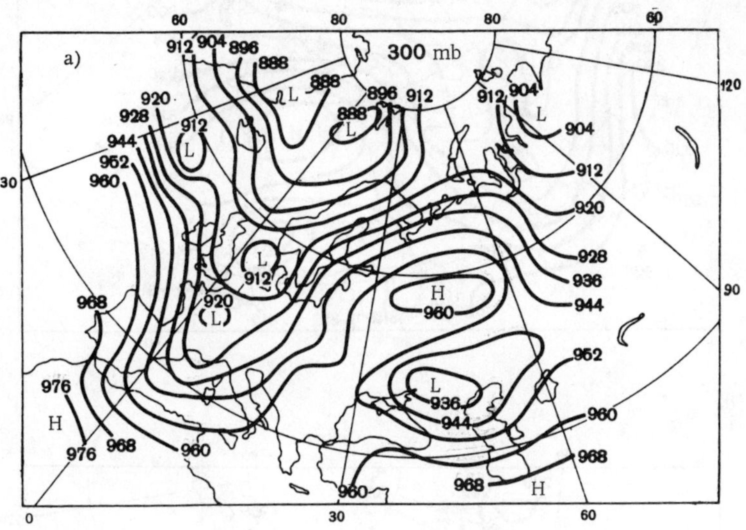

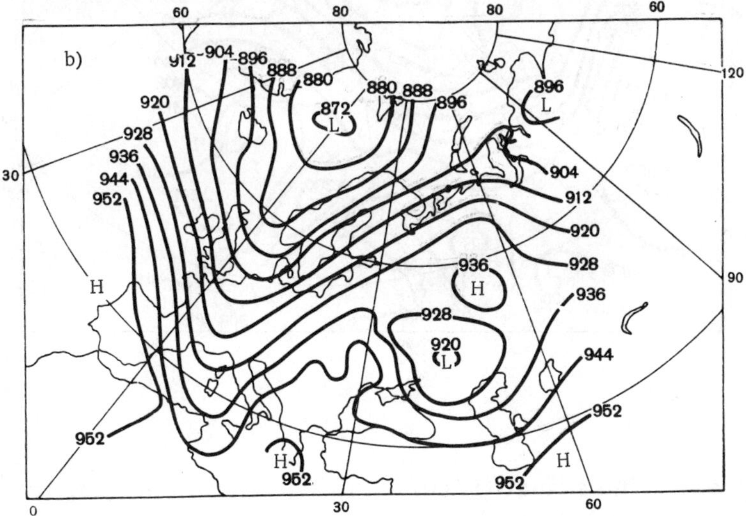

FIGURE 1.

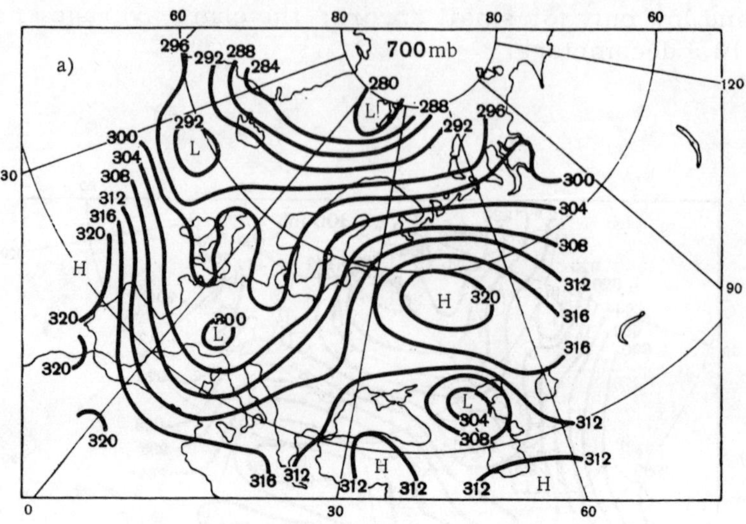

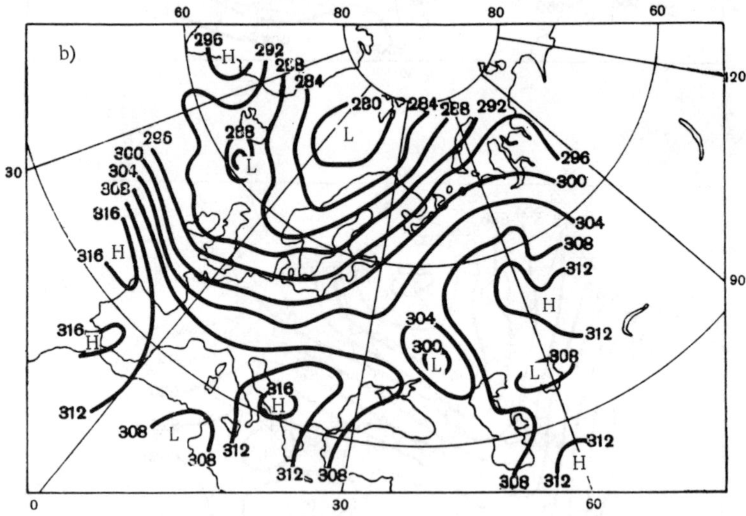

FIGURE 1.

We present the actual (Figure 1a) and calculated (Figure 1b) geopotential fields of the 300, 700 and 850 mb surfaces. The calculated fields reconstruct roughly many large-scale features of actual fields. It is noteworthy that, although the mean absolute error for the calculated 300-mb chart is 12.5 decameters, the qualitative pattern in this case is quite close to the actual pattern.

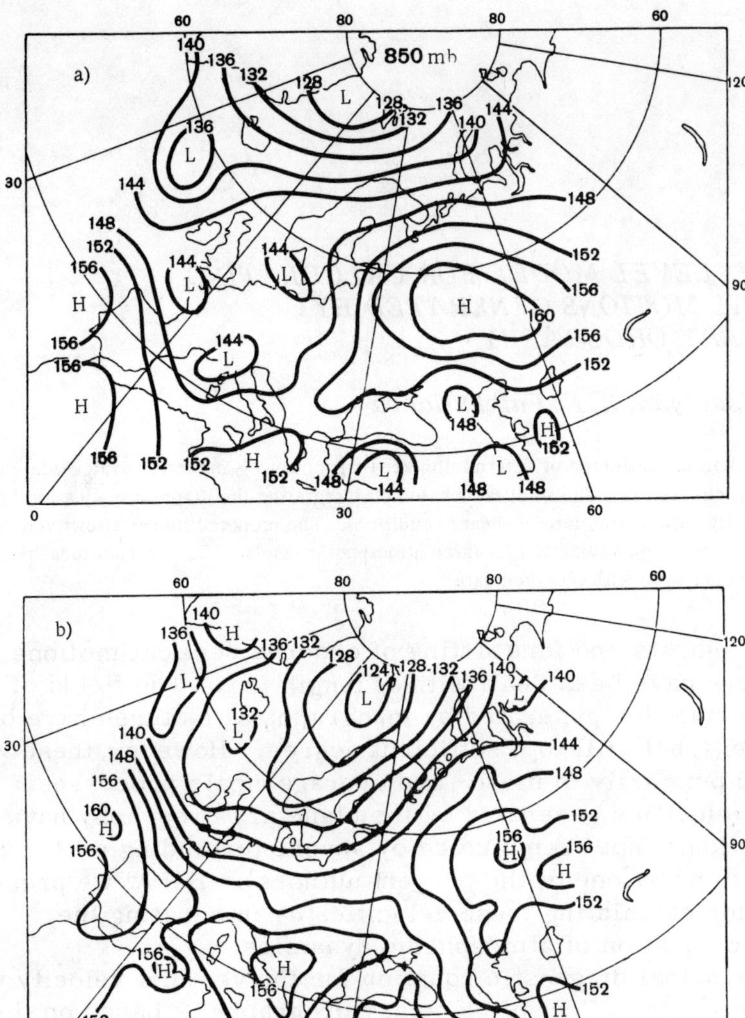

FIGURE 1.

BIBLIOGRAPHY

1. M u s a e l y a n , Sh.A. Nekotorye aspekty interpretatsii i ispol'zovaniya dannykh ob oblachnosti, poluchaemykh s meteorologicheskikh sputnikov (Some Aspects of the Interpretation and Utilization of Cloudiness Data Supplied by Meteorological Satellites).— DAN SSSR, No.5. 1963.

2. M c C l a i n , E.P. SINAP Problem: Present Status and Future Prospects — ESSA Techn. Rep. NESC, No.41, Washington D.C. Oct. 1967.

UDC 551.509

A THREE-LEVEL MODEL FOR CALCULATING VERTICAL MOTIONS GENERATED BY PLANETARY OROGRAPHY

Sh. A. Musaelyan, Z. N. Fatkhullaeva

The problem is considered of determining vertical motions generated by large-scale orography in the western circulation. The solution is based on integrating the diagnostic equation for the vertical velocity with appropriate boundary conditions. The proposed model allows vertical orographic motions to be calculated for three atmospheric levels. The computational results are analyzed and compared with observed data.

The diagnosis and forecasting of ordered vertical motions in the atmosphere have been dealt with at length /6/. The fields of vertical velocities generated by topographical features have been studied less, but still to a sufficient degree. However, these works pertained primarily to meso- and average-scale processes. The vertical velocities generated by the planetary orography have, as far as we know, not been studied by anyone (excluding preliminary investigations by one of the present authors). Below we present a model for calculating these velocities by integrating the diagnostic equation of atmospheric dynamics.

The principal diagnostic equation for the vertical velocity with allowance for the effect of planetary orography is based on the vorticity equation (more precisely, on the equation of the vertical vorticity component) and on the heat flux equation expressed in spherical coordinates. The first of these equations has the form

$$\frac{\partial \Omega}{\partial t} + \frac{v_\theta}{a} \frac{\partial}{\partial \theta}(\Omega + 2\omega \cos\theta) + \frac{v_\lambda}{a \sin\theta} \frac{\partial \Omega}{\partial \lambda} - \frac{\Omega + 2\omega \cos\theta}{\rho} \frac{\partial(\rho v_z)}{\partial z} -$$

$$- \frac{\nu}{a^2} \Delta\Omega = 0, \tag{1}$$

where λ is the longitude; $\theta = \dfrac{\pi}{2} - \varphi$; φ is the latitude; z is the vertical coordinate; v_θ, v_λ and v_z are the velocity components, respectively, along the θ, λ and z axes; a is the average radius of the earth $(a = 6.37 \cdot 10^6\,\text{m})$; ν is the horizontal mixing coefficient;

ρ is the density; t is the time; Δ and Ω are the Laplacian operator and the vertical component of the vorticity, respectively, given by

$$\Delta \equiv \frac{1}{\sin\theta}\left[\frac{\partial}{\partial\theta}\left(\sin\theta\,\frac{\partial}{\partial\theta}\right) + \frac{1}{\sin^2\theta}\,\frac{\partial^2}{\partial\lambda^2}\right], \tag{2}$$

$$\Omega = \frac{1}{a\sin\theta}\left[\frac{\partial}{\partial\theta}\,(v_\lambda\sin\theta) - \frac{\partial v_\theta}{\partial\lambda}\right]. \tag{2'}$$

Following Blinova /2/, it is assumed that the atmospheric circulation consists of a principal, purely western circulation and of small azonal disturbances superimposed on the latter. This assumption is based on analysis of a large volume of observational data and is in agreement with the principal tenets of the method of small perturbations and allows linearization of equation (1). It is assumed that the zonal components are functions of latitude and altitude:

$$v_\theta = v_\theta',\quad v_\lambda = \overline{v}_\lambda(\theta,\ z) + v_\lambda'(\theta,\ \lambda,\ z,\ t),$$
$$\Omega = \overline{\Omega}(\theta,\ z) + \Omega'(\theta,\ \lambda,\ z,\ t),\quad w = w'. \tag{3}$$

According to Blinova /2/, $\overline{v}_\lambda$ can be expressed by the following formula:

$$\overline{v}_\lambda = \alpha(z)\,a\sin\theta, \tag{4}$$

where α is the zonal atmospheric circulation index. It can be shown that

$$\overline{\Omega} = 2\alpha\cos\theta. \tag{5}$$

In notation (3) the azonal unsteady disturbances of meteorological characteristics of the state of the atmosphere, designated by corresponding symbols with primes, are taken to be quantities of the first order of smallness. Equation (1) is a nonlinear partial differential equation, and can be linearized by substituting expressions (3) into equation (1) and neglecting second-order infinitesimals (squares of primed functions or their products). Equation (1) then assumes the form

$$\frac{\partial\Omega'}{\partial t} - 2(\alpha+\omega)\sin\theta\,\frac{v_\theta'}{a} + \alpha\frac{\partial\Omega'}{\partial\lambda} - \frac{\nu}{a^2}\,\Delta\Omega' =$$

$$= \frac{2(\alpha+\omega)\cos\theta}{\overline{\rho}}\,\frac{\partial(\overline{\rho}v_z)}{\partial z}, \tag{6}$$

where the density is replaced by its standard value $\bar{\rho}(z)$. This simplification is utilized in all studies of the dynamics of large-scale atmospheric processes, since the density is then assumed to be independent of the horizontal coordinates. The planetary orography is now introduced in the form

$$z = \xi(\theta, \lambda). \tag{7}$$

During western circulation past terrestrial irregularities the velocity components v_λ' and v_θ' can be expressed by means of the following approximate formulas /7/:

$$v_\lambda' = \frac{1}{a}\frac{\partial \psi'}{\partial \theta} + \alpha_\xi a \eta \sin \theta, \tag{8}$$

$$v_\theta' = -\frac{1}{a \sin \theta}\frac{\partial \psi'}{\partial \lambda}, \tag{8'}$$

where $\eta(\theta, \lambda) = \dfrac{\xi(\theta, \lambda)}{H}$ (here $H = \dfrac{RT_0}{g} = 8 \cdot 10^3$ m is the height of the homogeneous atmosphere, R is the gas constant, g is the gravitational acceleration, and T_0 is the temperature); α_ξ is the circulation index at the mountain level. It is clear that in the absence of mountains $\eta \equiv 0$ and equations (8) and (8') reduce to ordinary relationships between the horizontal velocity components and the stream function (in this case as applied to azonal parts of these meteorological characteristics of the state of the atmosphere). Rigorously speaking, equations (8) and (8') are valid only for the middle levels of the atmosphere, but below they will be regarded as approximately valid for the entire atmosphere. We return now to equation (6) and transform it with allowance for the effect of orography, i. e., with consideration of equations (8) and (8').

On the basis of equation (2')

$$\Omega' = \frac{1}{a \sin \theta}\left[\frac{\partial}{\partial \theta}\left(v_\lambda' \sin \theta\right) - \frac{\partial v_\theta'}{\partial \lambda}\right],$$

and substitution of expressions (8) and (8') yields

$$\Omega' = \frac{1}{a^2 \sin \theta}\left[\frac{\partial}{\partial \theta}\left(\sin \theta \frac{\partial \psi'}{\partial \theta}\right) + \frac{1}{\sin \theta}\frac{\partial^2 \psi'}{\partial \lambda^2}\right] + \frac{\alpha_\xi}{\sin \theta}\frac{\partial}{\partial \theta}\left(\eta \sin^2 \theta\right),$$

or (introducing notation (2))

$$\Omega' = \frac{1}{a^2}\Delta \psi' + \frac{\alpha_\xi}{\sin \theta}\frac{\partial}{\partial \theta}\left(\eta \sin^2 \theta\right).$$

Consequently

$$\frac{\partial \Omega'}{\partial t} = \frac{1}{a^2} \frac{\partial \Delta \psi'}{\partial t},$$

$$\frac{\partial \Omega'}{\partial \lambda} = \frac{1}{a^2} \frac{\partial \Delta \psi'}{\partial \lambda} + \frac{\alpha_\xi}{\sin \theta} \frac{\partial^2}{\partial \theta \, \partial \lambda} (\eta \sin^2 \theta),$$

$$\Delta \Omega' = \frac{1}{a^2} \Delta \Delta \psi' + \alpha_\xi \Delta \left[\frac{1}{\sin \theta} \frac{\partial}{\partial \theta} (\eta \sin^2 \theta) \right]. \tag{9}$$

Equations (9) and (8') can be used to easily reduce equation (6) to the form

$$\frac{\partial \Delta \psi'}{\partial t} + \alpha \frac{\partial \Delta \psi'}{\partial \lambda} + 2\omega \frac{\partial \psi'}{\partial \lambda} - \frac{\nu}{a^2} \Delta \Delta \psi' = \frac{2a^2 \omega \cos \theta}{\bar{\rho}} \frac{\partial (\bar{\rho} v_z)}{\partial z} -$$

$$- 2\alpha \alpha_\xi a^2 \cos \theta \frac{\partial \eta}{\partial \lambda} - \alpha \alpha_\xi a^2 \sin \theta \frac{\partial^2 \eta}{\partial \theta \, \partial \lambda} + \nu \alpha \Delta \left[\frac{1}{\sin \theta} \frac{\partial}{\partial \theta} (\eta \sin^2 \theta) \right]. \tag{10}$$

In this study we do not intend to study the role of eddy friction as a dissipative factor. Hence, setting $\nu \equiv 0$ in equation (10), we obtain

$$\frac{\partial \Delta \psi'}{\partial t} + \alpha \frac{\partial \Delta \psi'}{\partial \lambda} + 2\omega \frac{\partial \psi'}{\partial \lambda} = \frac{2a^2 \omega \cos \theta}{\bar{\rho}} \frac{\partial (\bar{\rho} v_z)}{\partial z} -$$

$$- 2\alpha \alpha_\xi a^2 \cos \theta \frac{\partial \eta}{\partial \lambda} - \alpha \alpha_\xi a^2 \sin \theta \frac{\partial^2 \eta}{\partial \theta \, \partial \lambda}.$$

We now introduce the notation

$$\Gamma (\theta, \ \lambda, \ \zeta) = 2\alpha \alpha_\xi a^2 \cos \theta \frac{\partial \eta}{\partial \lambda} + \alpha^2 a^2 \sin \theta \frac{\partial^2 \eta}{\partial \theta \, \partial \lambda} \tag{11}$$

in which case the latter equation assumes the form

$$\frac{\partial \Delta \psi'}{\partial t} + \alpha (\zeta) \frac{\partial \Delta \psi'}{\partial \lambda} + 2\omega \frac{\partial \psi'}{\partial \lambda} - \Gamma = - \frac{2a^2 \omega g \cos \theta}{P} \frac{\partial (\bar{\rho} v_z)}{\partial \zeta}, \tag{11'}$$

where $\zeta = \dfrac{p}{P}$, p is the pressure, and P is the standard pressure at sea level.

It is best to take the heat influx equation in the form presented by Blinova in 1955 in her work on long-term forecasting of meteorological elements in a baroclinic atmosphere with allowance for vertical velocities and temperature stratification, and later in /5/:

$$\frac{\partial^2 \psi'}{\partial t \, \partial \zeta} + \alpha (\zeta) \frac{\partial^2 \psi'}{\partial \lambda \, \partial \zeta} - \frac{d\alpha}{d\zeta} \frac{\partial \psi'}{\partial \lambda} = \frac{R}{2\omega \zeta} \frac{\gamma_a - \gamma}{\cos \theta} v_z + \varepsilon, \tag{12}$$

where

$$\varepsilon = -\frac{R}{2\omega\zeta\cos\theta}\frac{1}{c_p\rho}(\varepsilon_1 + \varepsilon_2 + \varepsilon_3),$$

while ε_1, ε_2 and ε_3 are heat fluxes due to radiation, turbulence and phase transitions, respectively. It now remains to eliminate the time derivatives from equations (11') and (12) (see, for example, /4/). For this we apply the Laplacian operator to both sides of the last equation and subtract from it the derivative of the first equation with respect to ζ:

$$\frac{\partial^2}{\partial\zeta^2}\left(\frac{\bar\rho v_z}{\cos\theta}\right) + \frac{PR\,(\gamma_a - \gamma)}{4a^2\omega^2\bar\rho g\zeta\cos\theta}\Delta\left(\frac{\bar\rho v_z}{\cos^2\theta}\right) = \frac{P}{2a^2\omega g\cos^2\theta}\left[\frac{\partial\Gamma}{\partial\zeta} - 2\left(\frac{d\alpha}{d\zeta}\frac{\partial\Delta\psi'}{\partial\lambda} + \omega\frac{\partial^2\psi'}{\partial\lambda\,\partial\zeta}\right)\right] - \frac{P}{2a^2\omega g\cos^2\theta}\Delta\varepsilon. \tag{13}$$

The right-hand part of this equation contains three perturbation terms: the first term

$$\frac{P}{2a^2\omega g\cos^2\theta}\frac{\partial\Gamma}{\partial\zeta} = \overline{f}\,(\theta,\,\lambda,\,\zeta) \tag{14}$$

describes the contribution of orography; the second term

$$-\frac{2P}{2a^2\omega g\cos^2\theta}\left(\frac{d\alpha}{d\zeta}\frac{\partial\Delta\psi'}{\partial\lambda} + \omega\frac{\partial^2\psi'}{\partial\lambda\,\partial\zeta}\right) = \chi\,(\theta,\,\lambda,\,\zeta) \tag{15}$$

describes the contribution of the total pressure field (which is substantially simplified due to linearization); and finally the third term

$$-\frac{P}{2a^2\omega g\cos^2\theta}\Delta\varepsilon = \varphi\,(\theta,\,\lambda,\,\zeta) \tag{16}$$

describes the contribution of the heat flux.

Determination of the orographic vertical velocities thus reduces to integration of the equation

$$\frac{\partial^2 w}{\partial\zeta^2} + \sigma\,(\zeta)\,\Delta w = \overline{f}, \tag{17}$$

subject to boundary conditions:

$$w = \frac{\bar{p}}{\cos\theta}\,\alpha_\xi\,\frac{\partial\xi}{\partial\lambda} \quad \text{for} \quad \xi = 1, \tag{17'}$$

$$w = 0 \quad\quad\quad \text{for} \quad \xi = 0, \tag{17''}$$

where

$$w = \frac{\bar{p}v_z}{\cos\theta}\,,$$

$$\sigma = \frac{PR\,(\gamma_a - \gamma)}{4a^2\omega^2\bar{p}g\zeta\cos^2\theta}\,. \tag{18}$$

This problem is treated below in detail.

The determination of vertical velocities induced by nonuniform distribution of heat fluxes reduces to solving the equation

$$\frac{\partial^2 w}{\partial\zeta^2} + \bar{\sigma}\,(\zeta)\,\Delta w = \varphi \tag{19}$$

subject to boundary conditions:

$$w = 0 \quad \text{for} \quad \zeta = 1,$$
$$w = 0 \quad \text{for} \quad \zeta = 0.$$

Although this problem is stated completely, we shall not consider it here since it lies outside the scope of this paper. Its solution is similar to that of equation (17), which is considered below.

We now describe the scheme for calculating orographic vertical velocities for three atmospheric levels. The atmosphere is sub-divided into a number of layers with ζ equal to 0, 0.75, 0.5, 0.25 and 1. The second derivative of w with respect to the vertical coordinate at the different levels is approximated as follows:

$$\left(\frac{\partial^2 w}{\partial\zeta^2}\right)_{0.75} \approx 16\,(w_1 - 2w_{0.75} + w_{0.5}), \tag{20}$$

$$\left(\frac{\partial^2 w}{\partial\zeta^2}\right)_{0.5} \approx 16\,(w_{0.75} - 2w_{0.5} + w_{0.25}), \tag{20'}$$

$$\left(\frac{\partial^2 w}{\partial\zeta^2}\right)_{0.25} \approx 16\,(w_{0.5} - 2w_{0.25}). \tag{20''}$$

Relationship (20'') was derived on the basis of boundary condition (17''). Since we shall below write equation (17) for $\zeta = 0.75$, 0.5

and 0.25, it is expedient to first transform its right-hand side to these levels as well as the coefficient of the Laplacian operator on the left-hand side. For this we introduce the notation

$$\bar{\sigma} = \frac{PR\,(\gamma_a - \gamma)}{4a^2\omega^2 g\,\cos\theta}\,.\tag{21}$$

Thus

$$\sigma\,(\zeta) = \frac{\bar{\sigma}}{\rho\zeta}\,.\tag{22}$$

Quantity $\gamma = -\dfrac{\partial T}{\partial z}$ contained in the expression for σ depends in general on the altitude and plays an important role in a number of mesometeorological processes. Thus it was shown in /8/ for a flat earth that for a stably stratified atmosphere $(\gamma_a - \gamma > 0)$ the solution of the problem of orographic disturbances has a wavelike nature. In cases of unstable $(\gamma_a - \gamma < 0)$ or neutral $(\gamma_a - \gamma = 0)$ atmospheric stratification, wave processes up- and downstream of the obstacle cannot exit. Below, having reference to our study of large-scale atmospheric processes, we shall consider only the case of stable stratification. For convenience we introduce the following notation:

$$\sigma_{0.75} = \frac{\bar{\sigma}}{16 \cdot 0.75\bar{\rho}\,(0.75)}\,,$$

$$\sigma_{0.5} = \frac{\bar{\sigma}}{8\bar{\rho}\,(0.5)}\,,$$

$$\sigma_{0.25} = \frac{\bar{\sigma}}{4\bar{\rho}\,(0.25)}\,.\tag{23}$$

Consider now equation (11). The first term on the right-hand side describes sufficiently well the effect of large-scale orography, particularly when reference is had to meridionally aligned long ranges, such as the majority of mountain systems of the globe (Rocky Mountains, Andes, Urals, Scandinavian ranges, Caucasus, etc.). For such mountain systems the second term on the right-hand side of equation (11) is small and can be neglected.*

---

* Retention of the second term on the right-hand side does not involve any substantial difficulties. It would necessitate application of the Legendre polynomial relationship.

$$\frac{dP_n^m}{d\theta}\,\sin\theta = \frac{n\,(n - m + 1)}{2n + 1}\,P_{n+1}^m - \frac{(n + 1)\,(n + m)}{2n + 1}\,P_{n-1}^m\,.$$

The first term on the right-hand side of this equation describes satisfactorily the effect of mountain systems such as the Tibetan plateau, although for this case it would be better to retain both terms on the right-hand side of equation (11). With reference to the above, the second term on the right-hand side of equation (11) was omitted.

Hence

$$\frac{\partial \Gamma}{\partial \zeta} = 2 \frac{d\alpha}{d\zeta} \, \alpha_\xi a^2 \cos\theta \, \frac{\partial \eta}{\partial \lambda}.$$

Therefore,

$$\overline{f}(\theta,\, \lambda,\, \zeta) = \frac{P}{g} \, \frac{\alpha_\xi}{\omega} \, \frac{d\alpha}{d\zeta} \, \frac{\partial}{\partial \lambda}\left(\frac{\eta}{\cos\theta}\right). \tag{24}$$

Equation (17) is now expressed for levels $\zeta = 0.75$, $\zeta = 0.5$ and $\zeta = 0.25$:

$$-2w_{0.75} + w_{0.5} + \sigma_{0.75}\,\Delta w_{0.75} = f_{0.75} - w_1,$$
$$w_{0.75} - 2w_{0.5} + w_{0.25} + \sigma_{0.5}\,\Delta w_{0.5} = f_{0.5},$$
$$w_{0.5} - 2w_{0.25} + \sigma_{0.25}\,\Delta w_{0.25} = f_{0.25}, \tag{25}$$

where

$$f_{0.75} = \frac{\overline{f}_{0.75}}{16},$$

$$f_{0.5} = \frac{\overline{f}_{0.5}}{16},$$

$$f_{0.25} = \frac{\overline{f}_{0.75}}{16}.$$

For convenience we introduce the notation

$$\delta = \frac{P}{16g} \, \frac{\alpha_\xi}{\omega} \, \frac{d\alpha}{d\zeta},$$

$$\delta_1 = \frac{P}{16g} \, \frac{\alpha_\xi}{\omega} \left(\frac{d\alpha}{d\zeta}\right)_{0.75},$$

$$f_{0.75} = \delta_1 \frac{\partial}{\partial \lambda} \frac{\eta}{\cos\theta},$$

$$\delta_2 = \frac{P}{16g} \, \frac{\alpha_\xi}{\omega} \left(\frac{d\alpha}{d\zeta}\right)_{0.5},$$

$$f_{0.5} = \delta_2 \frac{\partial}{\partial \lambda} \frac{\eta}{\cos\theta},$$

$$\delta_3 = \frac{P}{16g}\frac{\varkappa_\xi}{\omega}\left(\frac{da}{d\zeta}\right)_{0,25},$$

$$f_{0,25} = \delta_3 \frac{\partial}{\partial\lambda}\frac{\eta}{\cos\theta}. \tag{26}$$

If function $\dfrac{\eta}{\cos\theta}$ is expressed as a series of spherical harmonics,

$$\frac{\eta}{\cos\theta} = \sum_n\sum_m \left(\eta_n^m\cos m\lambda + \eta_n^{'m}\sin m\lambda\right)P_n^m(\theta),$$

then (26) and (17'') yield

$$f_{0.75} = \delta_1\sum_n\sum_m m\left(\eta_n^{'m}\cos m\lambda - \eta_n^m\sin m\lambda\right)P_n^m,$$

$$f_{0,5} = \delta_2\sum_n\sum_m m\left(\eta_n^{'m}\cos m\lambda - \eta_n^m\sin m\lambda\right)P_n^m,$$

$$f_{0.25} = \delta_3\sum_n\sum_m m\left(\eta_n^{'m}\cos m\lambda - \eta_n^m\sin m\lambda\right)P_n^m,$$

$$w_1 = aH\bar\rho\sum_n\sum_m m\left(\eta_n^{'m}\cos m\lambda - \eta_n^m\sin m\lambda\right)P_n^m. \tag{27}$$

The solution of equations (25) is also sought in the form of a series of spherical harmonics:

$$w_{0.75} = \sum_n\sum_m \left(w_{1n}^m\cos m\lambda + w_{1n}^{'m}\sin m\lambda\right)P_n^m,$$

$$w_{0.5} = \sum_n\sum_m \left(w_{2n}^m\cos m\lambda + w_{2n}^{'m}\sin m\lambda\right)P_n^m,$$

$$w_{0.25} = \sum_n\sum_m \left(w_{3n}^m\cos m\lambda + w_{3n}^{'m}\sin m\lambda\right)P_n^m. \tag{28}$$

Substitution of (27) and (28) into equations (25) yields two systems of algebraic equations with which to determine the required coefficients w_{1n}^m, w_{2n}^m, w_{3n}^m and $w_{1n}^{'m}$, $w_{2n}^{'m}$, $w_{3n}^{'m}$.

Thus, equating the coefficients of the cosines gives the following system of equations for w_{1n}^m, w_{2n}^m and w_{3n}^m:

$$-[2 + n(n+1)\sigma_{0.75}]\,w_{1n}^m + w_{2n}^m = \delta_1 m\eta_n^{'m} - aH\bar\rho h_n^{'m},$$

$$w_{1n}^m - [2 + n(n+1)\sigma_{0.5}]\,w_{2n}^m + w_{3n}^m = \delta_2 m\eta_n^{'m},$$

$$w_{2n}^m - [2 + n(n+1)\sigma_{0.25}]\,w_{3n}^m = \delta_3 m\eta_n^{'m}.$$

The system of equations for coefficients w'^m_{1n}, w'^m_{2n} and w'^m_{3n} is constructed similarly. The required solutions of the problem at hand are now constructed with the aid of expressions (28).

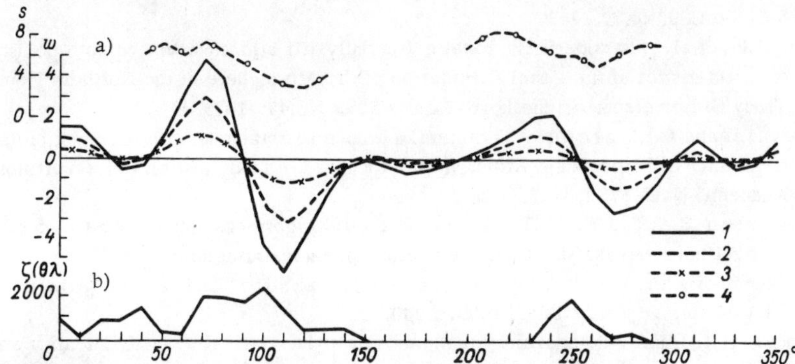

FIGURE 1. Orographic vertical velocities for levels corresponding to various values of ζ (a), and the profile of the terrestrial surface along 40°N (b):

$1 - \zeta = 0.75$; $2 - \zeta = 0.5$; $3 - \zeta = 0.25$; 4 — climatic values of the total cloud amount.

In conclusion we consider the following computational example. Figure 1a shows orographic vertical velocities along 40°N calculated from the above scheme for three atmospheric levels in winter. According to /3/, the climatic value of the circulation index for January was taken as $\alpha = 38.9 \cdot 10^{-3} \omega$. The same figure shows the profile of climatic values of the total cloud amount for the same month and for the same latitude, taken from /9/.

Figure 1b represents the corresponding terrestrial surface profile. These figures show that the generated orographic vertical velocities propagate downstream in a wavelike manner. Their amplitude decreases with horizontal as well as vertical distance from the obstacle. When the ascending motions attain their minimum at the windward side of mountains, the descending motions attain a maximum at their lee side.

The cloudiness profiles presented in Figure 1a are in sufficient agreement with the vertical velocity distribution. This shows that orographic vertical velocities play a paramount role in cloud generation processes.

BIBLIOGRAPHY

1. B e r l y a n d , T.G. Raspredelenie solnechnoi radiatsii na kontinentakh (Distribution of Solar Radiation on Continents). Leningrad, Gidrometeoizdat. 1961.
2. B l i n o v a , E.N. Gidrodinamicheskaya teoriya voln davleniya i tsentrov deistviya atmosfery (Hydrodynamic Theory of Pressure Waves and Atmospheric Activity Centers).— DAN SSSR, Vol.39, No.7. 1943.
3. B o i k o , A.P., et al. Kharakteristiki zonal'noi tsirkulyatsii atmosfery v severnom polusharii (Characteristics of the Zonal Circulation of the Atmosphere in the Northern Hemisphere).— Trudy Gidrometeorologicheskogo Tsentra SSSR, No.47. 1970.
4. B u l e e v , I.I. and G.I.M a r c h u k . O dinamike krupnomasshtabnykh atmosfernykh protsessov (Dynamics of Large-Scale Atmospheric Processes).— Trudy Instituta Fiziki Atmosfery Akademii Nauk SSSR, No.2. 1958.
5. M a s h k o v i c h , S.A. and Y a.M.K h e i f e t s . K teorii dolgosrochnogo prognoza s uchetom vertikal'noi stratifikatsii atmosfery i turbulentnogo peremeshivaniya (Theory of Long-Range Forecasting with Allowance for the Vertical Stratification of the Atmosphere and Eddy Mixing).— Trudy TsIP, No.93. 1960.
6. M o r s k o i , G .I. Analiz issledovanii krupnomasshtabnykh vertikal'nykh dvizhenii v atmosfere (Analysis of Studies of Large-Scale Vertical Motions in the Atmosphere).— Trudy Gidrometeorologicheskogo Tsentra SSSR, No.36. 1968.
7. M u s a e l y a n , Sh.A. Planetarnye orograficheskie volny v zapadnom potoke (Planetary Orographic Waves in the Western Circulation).— In: Sbornik "Voprosy dinamicheskoi meteorologii." Izdatel'stvo AN SSSR. 1960.
8. M u s a e l y a n , Sh.A. Volny prepyatstvii v atmosfere (Obstacle Waves in the Atmosphere). Leningrad, Gidrometeoizdat. 1962.
9. M u s a e l y a n , Sh.A. Klimaticheskie dannye ob oblachnom pokrove Zemli i perspektivy ikh ispol'zovaniya pri reshenii zadach sputnikovoi meteorologii (Climatic Data on the Cloud Cover of the Earth and Prospects of their Utilization in Solving Problems of Satellite Meteorology).— Trudy Gidrometeorologicheskogo Tsentra SSSR, No.30. 1968.

UDC 551.509

THE CLOUD FIELD AND CYCLOGENESIS

T. P. Popova, T. A. Irisova

This paper deals with photographs of the cloud cover during cyclogenesis over the Mediterranean and arrival of cyclones over European USSR, as well as inception of cyclones at southern Scandinavian shores. It is shown that cyclogenesis is related to the formation of characteristic cap clouds.

The study of cyclogenesis and determination of the time of cyclone inception remain the most timely and the most difficult meteorological problem. It is hence natural to attempt to find in the new, very refined information supplied by satellites new indications which point to the initiation of cyclogenesis.

This paper presents some results of such utilization of TV cloud-cover photographs from meteorological satellites. Three aspects are treated: secondary cloud vortices arising within extensive cyclones, and the associated cyclogenesis at the cold front; arrival of southern cyclones over European USSR and the associated cloud situation; and cloud fields during cyclogenesis at the southern tip of Scandinavia.

Secondary cloud vortices have been dealt with in the literature as a variety of cloud vortices and as an indicator of possible cyclogenesis. It was shown in /1, 2/ that when the secondary vortex arises in cold air past a front, a wave or cyclone can arise within several hours at the cold front. If the secondary cloud vortex arises in the forward or left part of the cyclone, there is no cyclogenesis. Calculations show that secondary cloud vortices form in that part of the pressure field where mesoscale cyclonic circulation arises, i. e., in locations where a closed cell of positive vorticity values is detected in the field of the Laplacian calculated from the surface pressure. Here the center of the cell coincides approximately with the center of the secondary cloud vortex. It should be remembered that the surface pressure field in these cases does not always have a clearly expressed trough profile. In a number of cases the pressure field is characterized only by a rapid

reduction in the gradient in the direction of low pressure. Such a
distribution usually exists at the cyclone side of the zone of strong
winds (to the left of its axis). Such a situation frequently arises at
the southern periphery of an old cyclone, when baric gradients decrease in its central part as a result of its filling and of rapid
pressure rise.

Being situated in the zone of strong winds, the center of positive
vorticity, and correspondingly the secondary cloud vortex, move
toward the cold front and under satisfactory conditions can promote
the formation of a wave or cyclone at the front.

In this article the secondary vortex is not treated independently,
but rather as a particular case for cyclogenesis at the southern
shores of Scandinavia.

CLOUD FIELD DURING CYCLOGENESIS IN SOUTHERN EUROPE

Much attention is paid to the conditions under which cyclones
arrive over European USSR, because cyclones forming at southern
boundaries of European USSR above the Black and Caspian seas,
and arriving directly over the USSR, induce sharp changes in the
weather (strong winds, precipitation, snow and sand storms, ice-
covered ground). Forecasting of the inception, development and
movement of these cyclones involves certain difficulties very much
related with features of cyclone formation under complex orographic
conditions characteristic of the southern boundary regions of
European USSR, and also with the insufficient quantity of meteorological and aerological observations over the Near East.

At present ground-observation data are supplemented to a large
extent by cloud data supplied from meteorological satellites.
Satellite information augments immensely our information of cloud
structure, and has introduced new concepts as to the structure of
medium and large-scale cloud systems. In this study satellite
cloud information is used as primary data. We utilized the stored
satellite information and cloud photographs obtained by the Meteor
cosmic meteorological system over 1.5 years (January 1967 to
July 1968). Cloud photographs from ESSA or Nimbus satellites
were used for experimental purposes.

The criterion for selecting data for analysis was the presence of
cyclonic activity over the Mediterranean, Asia Minor, Black and
Caspian seas. Maps were constructed of trajectories of cyclones
which formed in these regions. The purpose of constructing these

maps was to utilize these trajectories as an indicator in the cyclone
sampling system.   All the cyclones were subdivided into nine
groups, each of which had approximately the same trajectories.
The recurrence of cyclones moving along different trajectories is
given in Table 1.   Approximately two-thirds of the cyclones
originating over the Caspian Sea arrived at the Volga River region
and Southern Urals.   Less than half of the cyclones which originated
over the Mediterranean and the Balkans passed through the Black
Sea to European USSR.   Some cyclones from the Balkans arrived at
the extreme western regions of European USSR.   Cyclones arriving
in the USSR pass very infrequently through Asia Minor; the path
of such cyclones passed through Transcaucasia to the Caspian Sea.
A particular case is a cyclone which originated over the northern
part of the Apennines and passed through the Balkans to Asia Minor,
Transcaucasia, through the Caspian Sea, and arrived at the Southern
Urals and Western Siberia.   The cyclone was tracked from 1 to
9 March 1967.

TABLE 1. Cyclone trajectories and their recurrence from January 1967 to July 1968

Group	Region of inception	Direction of motion	Number of cyclones
1	Caspian Sea	Volga-River region, Southern Urals	13
2	The same	Slow-moving	8
3	Black Sea (northern part)	Central European USSR	3
4	Black Sea	Slow-moving	7
5	Mediterranean, Balkans	European USSR through the Black Sea	16
6	The same	Do not arrive at the Black Sea and European USSR	22
7	" "	European USSR (western regions)	5
8	Asia Minor	Caspian Sea by way of Transcaucasia	3
9	The same	Slow-moving	13
Special case	Mediterranean	Asia Minor, Caspian Sea, Southern Urals	1

   In addition to synoptic maps and aerological soundings, cloud
pictures obtained from meteorological satellites were available for
69 of the 91 cyclones listed in Table 1.   Hence the development of
the synoptic process and the evolution of the cloud cover could be
studied.   As a result we detected various features in the structure
of the cloud cover preceding the cyclone inception, and the evolution
of clouds was followed during the motion and development of these
cyclones.   Indicators were established with which one could analyze

the inception of cyclogenesis over the Mediterranean, Balkans and
Near East, as well as the probability that cyclones will appear over
Southern European USSR.   The hypotheses put forward in /3/ were
also verified.

The indication that cyclogenesis starts over these regions is the
formation of stratiformis in the form of a cap with an anticyclonically curved northern edge, frequently bordered by a shadow band.
Here the clouds have a band structure.   Sometimes the bands
had a very much curved lenticular shape.   Ground-observation data
indicated that these were primarily Ci, Cs, As, and Ac-As, which
are related to warm-air advection.   During the initial stage of
cyclogenesis the vortex structure of the cloud cover is not seen,
but appears as the cyclone develops and as cold air invades the
rear and central parts of the cyclone.

If a cap of stratiformis appears over the Balkans, Black and
Caspian seas, and Northern Caucasus, then cyclogenesis should be
expected in this region.   If there is no perceptible spiral structure
in the cloud cover here, one should expect the cyclone to appear
over European USSR.   If, on the other hand, the clouds have a
developed spiral structure, then the arrival of a cyclone over
European USSR is less probable.

One frequently observes splitting of the cyclone over the Eastern
Mediterranean area.   In these cases the old cyclone remains over
the Balkans or moves to Asia Minor.   The cloud field in it acquires
a spiral structure.   The new cyclone originates ahead of the old,
due to the formation of a characteristic cap cloud with anticyclonically bent northern boundary somewhat separated from the cloud
system of the old cyclone.   This cyclone usually appears over
European USSR.

In individual cases southern cyclones arose directly over the
USSR (Crimea and Southern Ukraine).   In these cases the inception
of the cyclone is also preceded by the formation of the cap cloud,
but usually of smaller dimensions.   This is also stratiformis with
a band structure and with an anticyclonically curved northern edge.
In the pressure field at the ground there exists an area with a
somewhat depressed pressure, but this is still not a cyclone.   The
cyclone appears 12—24 hours after the appearance of the cap cloud.

A somewhat separate group is that of cyclones arising over the
Carpathian Mountains: these appear over western European USSR
and move along the frontal zone, along the western periphery of a
strong anticyclone which occupies the major part of European USSR.

As an example, we analyze the inception of cyclones on 8—14
January 1967 and 21—22 April 1969.

Cyclone of 8—14 January 1967.   During this period a series of
three cyclones passed over the Mediterranean, Balkans, Black Sea

and southern European USSR. They originated over the western
half of the Mediterranean. Two of them passed through the Balkans
and the west of the Black Sea to Southern Ukraine and further,
moving northeastward, and arrived at the Northern Urals. The
third cyclone moved from the Mediterranean to Asia Minor and
then, after moving over the Caucasian Range, passed over the
Caspian Sea and reached the Aral Sea. Simultaneously with the
movement of these cyclones frequent cyclogenesis was observed
over the Caspian and Black seas. This cyclogenesis was weak and
local; it was due to the orography and was observed at leeward
slopes during the flow of air through the Caucasian Range and
mountains of Asia Minor.

On the morning of 8 January a small cyclone with one closed
1005-mb isobar was observed in the pressure depression
occupying the Mediterranean and extreme south of Europe against
the background of a ground pressure of about 1010 mb over the
Balkans. This cyclone originated over the north Apennines and
was related to an old front which limited the invasion of cold air
over the Mediterranean; this occurred some time ago. The
cyclone died out. Its clouds did not constitute a complete ordered
structure. Sufficiently dense clouds persisted in the eastern part
of the cyclone and along that part of the front which extended to
central European USSR.

During subsequent days this cyclone was regenerated as a result
of a new cold invasion over the Mediterranean. The cyclone was
delineated by two closed isobars and started moving rapidly toward
the Black Sea and then to European USSR. This regeneration of the
cyclone and its appearance over European USSR was preceded by
the formation of a stratiform cloud mass over the Carpathian
Mountains. Twenty-four hours later, on the morning of 9 January,
the cyclone was already over the Ukraine.

At the same time a wave, which moved rapidly eastward in the
direction of the Carpathian Mountains, formed at the new arctic
front above the western Mediterranean. A heavy cap cloud *AA*
(Figure 1) formed ahead of the wave, and this cap occupied the
entire Balkan Peninsula. The northern edge had anticyclonic
curvature. The clouds were predominantly stratiformis, high, with
clearly seen shadows along the northern edge of the cap. It is
characteristic that this cap was shifted eastward (ahead) relative
to the wave. Near the center of the wave cumuliformis was
observed, in addition to stratiformis. Subsequently the wave moved
in the direction of the cap cloud and formed into a cyclone, which
passed rapidly over the Balkans and arrived, by way of the Black
Sea, over European USSR.

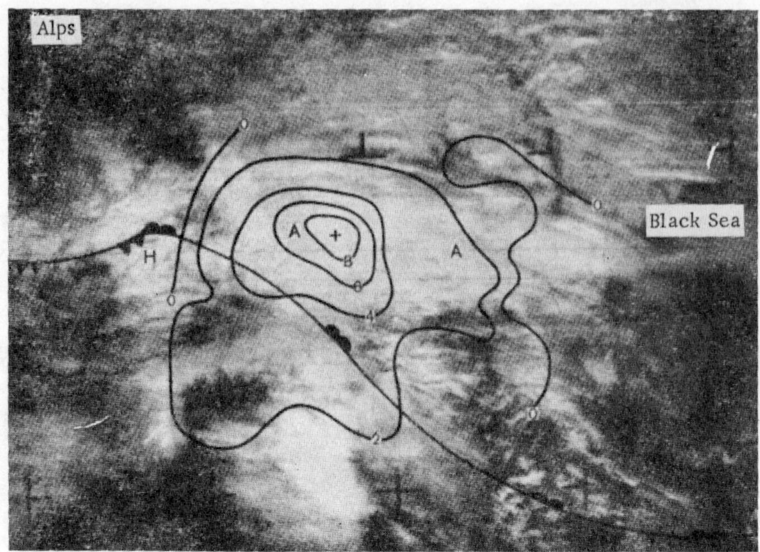

FIGURE 1. Cloudiness on 9 January 1967 (ESSA, 1236 hrs). Temperature
advection isolines at the 700-mb level (deg/12 hours).

Synoptic analysis showed that the tremendous stratiformis mass
in the form of a cap over the Balkans arose in the zone of intensive
head advection ahead of the frontal wave. Figure 1 shows temperature advection isolines at the 700-mb level at 1200 hrs on 9 January,
when the cloud cover was photographed from the Nimbus satellite.

Subsequently the cyclone continued to move rapidly eastward
(with a daily average velocity of 60 km/hr). On the morning of
10 January the cyclone center was located over the eastern Black
Sea, and 24 hours later over the Central Black-Earth Region of
European USSR. This, as the preceding cyclone, deepened
especially upon arriving over the Black Sea. The cloud system of
this cyclone had, on 10 January when it was situated over the Black
Sea, an elongated and somewhat deformed lenticular shape on the
cloud picture (Figure 2).

Simultaneously with the motion of the cyclone over European
USSR, high-cloud bands, elongated along air flows in the jet stream
which formed over Southern Europe, again appeared over the
western part of the Mediterranean. A cyclone in which occlusion
had already started moved to the Bay of Biscay from the Atlantic.
Subsequently this cyclone moved over the western Mediterranean,
and continued to occlude. The occlusion point on the morning of
11 January was already situated at the eastern periphery of the
cyclone, while the cloud system of the latter consisted of a vortex
with a clearly expressed spiral. The further eastward movement

of the cyclone was accompanied by its slight deepening and
deformation.   On the morning of 12 January two centers were
observed at the ground surface in the cyclone: over western shores
of Sicily (principal center) and above the south of the Adriatic Sea
(originated at the occlusion point).

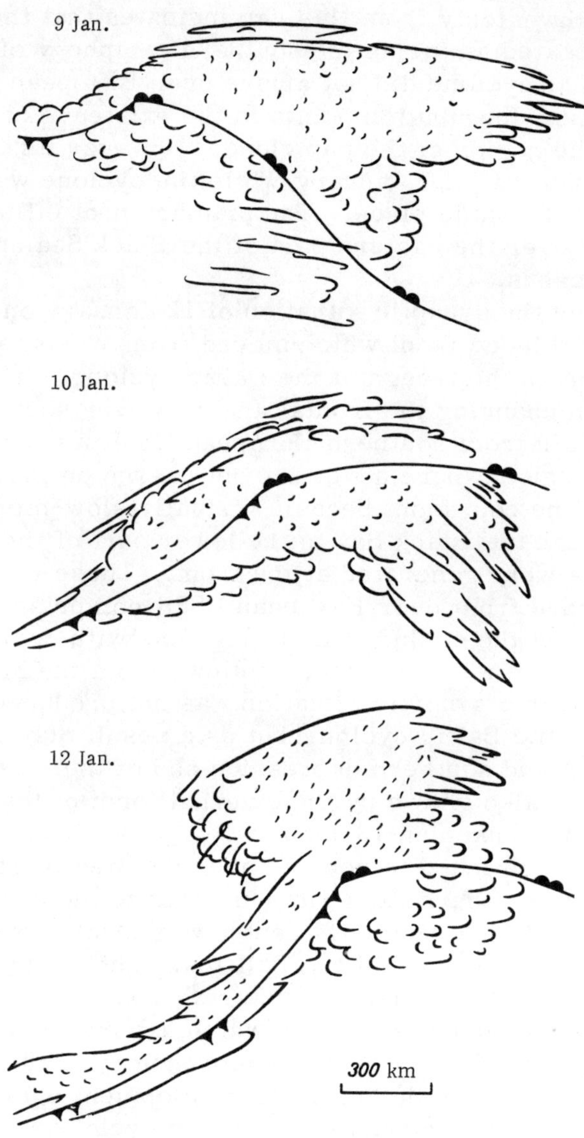

FIGURE 2. Schematic of cloud patterns of Mediterranean
cyclones on 9, 10 and 12 January 1967.

The cloud system of the cyclone underwent a major change. Cumulus prevailed near the principal center. The cloud cover structure had a somewhat spiral shape. These cloud forms are characteristic for an occluded, cold-air-filled cyclone. Near the occlusion point, in the region of the second cyclone [center] above the southern part of the Adriatic, the clouds are stratiformis. The cap cloud covering the Balkans was clearly seen. The cloud band extending southwesterly from this cap indicates that the cold invasion penetrated as far as the southern periphery of the cyclone (Figure 2). The cyclone did not arrive over European USSR.

The advection distribution points to the existence of warm-air advection in the region of the cap cloud. However its intensity is much lower than that of 9 January, while the cyclone was located in approximately the same place. The primary heat efflux was by way of Asia Minor over the eastern part of the Black Sea and the Northern Caucasus.

In examining the synoptic situation of 12 January one must consider the occluded front which moved from Western Europe to European USSR in the trough of the Baltic cyclone. It played a certain role in changing the frontal analysis over southern USSR.

The warm air from southern European USSR moved to the lower Volga-River region, where a warm front passed on the morning of 13 January. The cold front became virtually slow-moving, and extended through the Black Sea to the lower part of the Dnieper River. Waves were generated at the front. These fronts together with the occluded front over European USSR comprised a system similar to that of the occluded Baltic cyclone with occlusion point over eastern Ukraine. However, it follows from the history of the process that such a synoptic situation was not produced as a result of occlusion of the Baltic cyclone, but as a result of a combination of the northern and southern depression and of their associated fronts. This is also shown by the external form of the cloud cover on the photograph taken on 13 January.

The cyclone continued to occlude over the Mediterranean; it moved from the southern Adriatic somewhat to the southeast, and on the morning of 13 January its center was located over the Mediterranean to the south of the Balkans. The cloud system of this cyclone acquired a vortex structure. This terminated the cyclogenesis over southern Europe which extended over seven days.

It is found that differences in the morphology of the cloud cover of cyclones, which move during their development over the Black Sea and arrive over European USSR, and of cyclones that already have a tendency to stabilize over the Mediterranean basin or to move to Asia Minor are detected even before the appearance of a

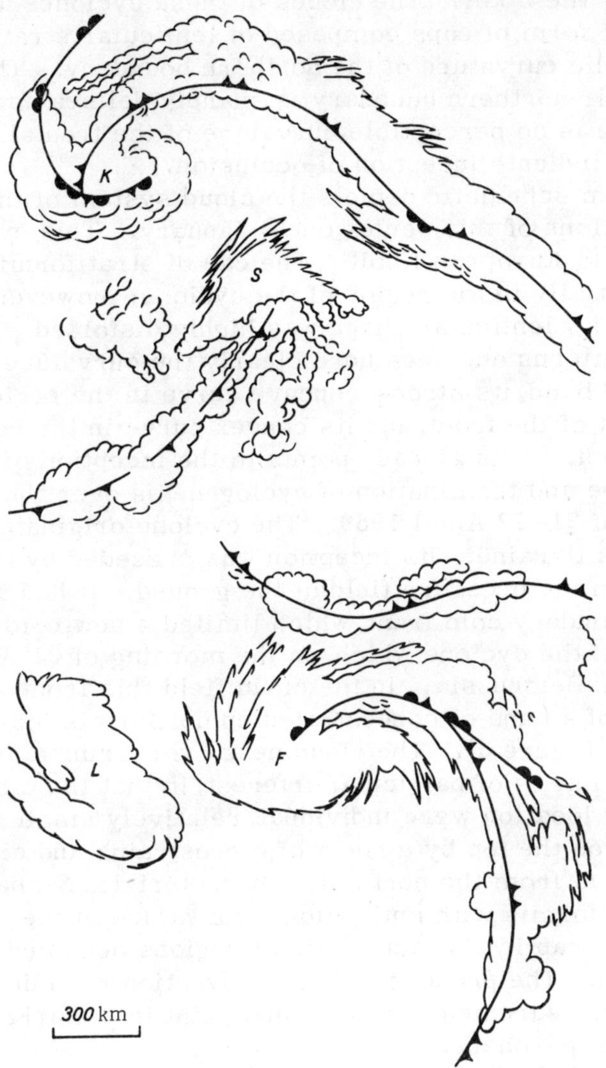

FIGURE 3. Schematic of cloud cover for 21—22 April 1969.

clearly defined vortex. Figure 2 presents schematics of cloud systems of two cyclones from the series considered above. The upper two schematics depict in sequence the cloudiness in the cyclone which originated over the Mediterranean on 9 January and arrived over the USSR. The clouds of these cyclones are represented in the form of caps composed of lenticular stratiformis with an anticyclonic curvature of the northern boundary. The clouds are high, and their northern boundary is sharply delineated by a shadow band. There is no perceptible curvature of the frontal cloud band which would indicate inception of occlusion.

The bottom schematic depicts the cloud system of the last, terminal cyclone of this series on 12 January. This cyclone did not move on to European USSR. The cap of stratiformis was still retained partially in the region of the cyclone; however, it is much smaller and its lenticular shape was highly distorted. Unlike the preceding patterns one sees here clearly the curvature of the frontal cloud band, its strong, concave curve in the region of the cold segment of the front, and its convex curve in the region of the warm segment. This already points to the inception of occlusion of the cyclone and termination of cyclogenesis over the Balkans.

Cyclone of 21–22 April 1969. The cyclone originated over southwestern Ukraine. Its inception was preceded by the formation of a trough in the pressure field at the ground. It had associated with it a secondary cold front, which limited a new cold-air invasion in the rear of the cyclone which, on the morning of 21 April, was situated over Belorussia. In the cloud field this front was detected in the form of a band of concentrated cumuliformis moving toward the Balkans (Figure 3). The cloudiness over Crimea and southwestern Ukraine is of particular interest, in that the clouds in the cyclogenesis location were individual, relatively small cumulonimbus covered on the top by a sheet of cirrostratus and cirrus. Adjoining them from the north is a characteristic S-shaped cap of banded stratiformis with anticyclonic curvature of the northern edge. Such a cap is characteristic of regions occupied by warm-air advection. The presence of such advection is indicated by radiosonde measurements in Kiev, and point to a marked right wind shift in the troposphere.

During the following day there developed a cyclone with a closed isobar which slowly (an average of 20 km/hr) moved to the northeast. On the morning of 22 April its center was located at Kursk. The cyclone produced prolonged precipitation in the form of snow and rain. Its cloud system by now consisted of a clearly delineated cloud spiral F. On the other hand cyclone K, located on the morning of 21 April over Belorussia, filled. The clouds within

it degenerated perceptibly and cloud spiral *L* became isolated.   The
cloud systems of the warm and cold fronts, with which the cloud
spiral was closely associated, moved northward and broke-up
markedly.   The cold air in the rear of the secondary front extended
over the entire Ukraine and Black Sea.   The secondary front thus
became a primary front.

This case is interesting in that the cyclone did not originate from
a wave at the front; it originated under conditions when synoptic
maps clearly depict the cold segment of the front, but no warm
front is found.   The latter formed as cold air moved into the rear
of the trough and as the advection of heat increased in the forward
part of the trough.

CLOUD FIELD DURING CYCLOGENESIS OVER
THE DANISH STRAITS AND THE NORTH SEA

The inception of cyclones over the Danish straits (Skagerrak,
Kattegat and Oresund) is one of the most frequent phenomena in
synoptic processes over Europe.   The frequency of these cyclones
is quite high and they determine to a large extent the nature of
weather over central and northern regions of the continent,
particularly during transition seasons.   Determination of the
inception of genesis of these cyclones and their evaluation present
a number of difficulties to the forecaster.   The feasibility of
utilizing satellite information can make a substantial contribution
to the solution of this problem.

The studies were carried out on the basis of available synoptic
material and analysis of cloud photographs for one year (April 1969
to March 1970).   We selected all cases of genesis or passing of
cyclones through the region of the Danish straits and southern
Scandinavia.   A total of 39 such cyclones were observed for the
year, half of them during fall and one-quarter in spring.   Very few
cyclones occurred in winter and summer (Table 2).   In 23 cases
they formed in the region of the Danish straits, at the southwestern
tip of the Scandinavian Peninsula and its adjoining part of the North
Sea, while 14 cyclones passed over this region from their breeding
grounds over the Atlantic or British Isles.   A cyclone once arose
directly over the Baltic Sea and on another occasion over the center
of Western Europe.   These cyclones, during their subsequent
development and movement over Northern Europe, behaved as
typical western cyclones.

TABLE 2. Frequency of cyclones by seasons

Region of cyclone inception	Spring	Summer	Fall	Winter	Total for the year
Southern Scandinavia, Danish straits ..	7	5	9	2	23
Norwegian Sea, Atlantic	–	1	4	1	6
British Isles	2	–	4	2	8
Baltic Sea	–	–	1	–	1
Western Europe	–	1	–	–	1
Total number of Baltic cyclones	9	7	18	5	39

The cloud fields of the first group of cyclones are studied here,
i. e., those which were generated at the southern Scandinavian
shores.   The trajectories of these cyclones are subdivided into
six principal types:

Filling region	Northern European USSR	Arctic Ocean	Scandi- navia	Baltic	Central European USSR and Central Urals	Southern European USSR and Urals
Number of cyclones	8	1	2	6	4	2

The predominant direction was northeastward: the cyclones
moved over the Baltic Sea, Karelia, and northern European USSR.
In exceptional cases such cyclones then moved to the Arctic Ocean
in the region of the Barents and Kara seas, where they regenerated
at the Arctic front and transformed into diving West Siberian
cyclones.   Some of the cyclones filled over Scandinavia and the
Baltic Sea, while some moved to central European USSR, terminat-
ing in the Cis-Ural region.   The above cyclones do not include the
numerous slow-moving and wide depressions, which result from
cyclonic activity over these regions.

It is known that the majority of cyclones arriving at the West
European coast are already old occluded systems, and they are
finally damped-out over the continent. Cyclogenesis over Central
Europe is either in the nature of regeneration of damping cyclones
or the inception of new ones. The cloud system of such cyclones
is affected by previously existing cloud systems. The development
of cyclones only from a wave at the cold or stationary front is
a very infrequent occurrence, and generally does not result in the
formation of a deep cyclone. Most frequently the evolution of such

a cyclone terminates at the young-cyclone stage, whereupon the cyclone does not develop further but starts to fill gradually.

Cyclogenesis is most frequent against the background of old filled occluded systems. Here one can distinguish two characteristic types of processes, and both assume the presence of a multi-centered depression over Northern Europe. The first type provides for deepening and development of one of the individual cyclonic centers, while the second involves the development of waves at a stationary or cold front passing through the depression. The development of a cyclone against the background of an old broken-up depression is frequently taken as a combination of both types.

The first of these types of synoptic process in the purest form was observed over Scandinavia at the end of November 1969. As a ryle, cyclones arising under these circumstances do not become too deep or too wide. Usually they are limited by two or three closed isobars; they do not last longer than two or three days. Having arisen in the trough of a multicentered depression, they again become a trough. Such a cyclone is observed in the cloud field as a small cloud vortex, observed against the background of the spotty cloudiness of the filled depression.

The second type of process is the most frequently encountered and can be divided into two subtypes: the first comprises two parallel systems of fronts with waves, a cyclone with a clearly defined warm sector being formed at the approaching tips; the second subtype comprises a wave at a stationary or cold front passing over the old breaking-up depression, provided that an old front, possibly an occlusion, forms in the warm sector of the wave.

Analyses of synoptic data and cloud photographs obtained from satellites have made it possible to single out a number of features in the cloud structure which precede the inception of a cyclone, and to follow the evolution of cloudiness during further cyclone development. The formation of a wave, at a front which is clearly seen in the form of widening of the frontal cloud band, still does not point to the possibility of its subsequent development. In order for a cyclone to develop, a large supply of cold air to the rear of the wave and the removal of heat from its front are needed. Only a clear advective pair, which results in increasing thermal and pressure gradients, produces conditions under which the wave can develop into a cyclone. The sign of inception of cyclogenesis is the formation of a stratiformis cap at the tip of the wave, pointing to the advection of warm air. This cap has a characteristic anti-cyclonic curvature of the northern edge, and is frequently separated by a shadow band from the frontal cloudiness at the tip of the wave. The cloud cover acquires a clearly expressed band structure.

Ground-collected data show that the stratiformis cap consists primarily of high and middle stratiformis (Cs, As). The vortex structure of the cloud cover cannot be distinguished either at the cap or at the tip of the wave. It appears with development of the cyclone during penetration of cold air into the rear and center of the cyclone; this is characteristic of the stage of maximum development and occlusion of cyclones.

Since the cyclones under consideration develop primarily against the background of old, filled depressions, their cloud cover has a complex structure, consisting of frontal band, cloud caps at the tip of the wave, and accumulation of clouds in the warm sector consisting of the remaining old cloud systems. Hence, in order to forecast cyclogenesis, it is important to single out the cloud system of the wave with its adjoining cloud cap against the background of this variegated cloud pattern.

The development of cyclones from a frontal wave at a secondary front makes up a somewhat separate group. Usually this also occurs in the region of an old, disintegrating cyclone. Inception of cyclonic activity becomes possible due to a large invasion of cold masses into the rear of the cyclone behind the secondary front. A cyclone still prevails at higher altitudes, at the ground it is almost filled, and only secondary fronts exist in it. The frontal zone of the secondary front is not very clearly expressed in the cloudiness field, and consists of spotty clouds, primarily cumuliformis with breaks. The wave arising at the front causes deformation of the frontal cloud band, its widening, and increased density at the tops. The wave develops into a young cyclone provided TV photographs show at its tip a clear b. ight spot, which serves as the primary indication of cyclogenesis at the secondary front.

Special mention should be made of cases of cyclogenesis during the combination of a secondary cloud vortex with a frontal wave, a process giving the impression of "instantaneous occlusion." On TV photographs the secondary cloud vortex is observed in the form of a bright white spot, consisting of dense cumuliformis twisted toward the center and having a shape similar to a comma, which is the name assigned it in meteorology. When the comma comes close to the frontal cloud band, a cloud field pattern is produced which is similar to an occlusion spiral. At the ground it has corresponding to it a low-level cyclone, limited by one primary and one intermediate isobar. The axis of the thermal ridge passes high above this cyclone. The high-pressure field also indicates the presence of a ridge above this region.

Examples of the cloud field during cyclogenesis at the southern tip of Scandinavia are given in Figure 4. On 1 November 1969 a cyclone was generated above the Skagerrak straits in the region of an old deep cyclone which occupied Northern Europe. A cloud cap, which appeared to be situated on the frontal cloudiness, the edge of which is delineated by a shadow band, formed above southern Scandinavia. The cap was formed of stratiformis (Ci, Cs, As), the diagram of which clearly shows bands parallel to the outer northern edge of the cap, which has an anticyclonic curvature. The cyclone, once generated, passed through all the development stages while moving from the Baltic Sea to the Northern Urals.

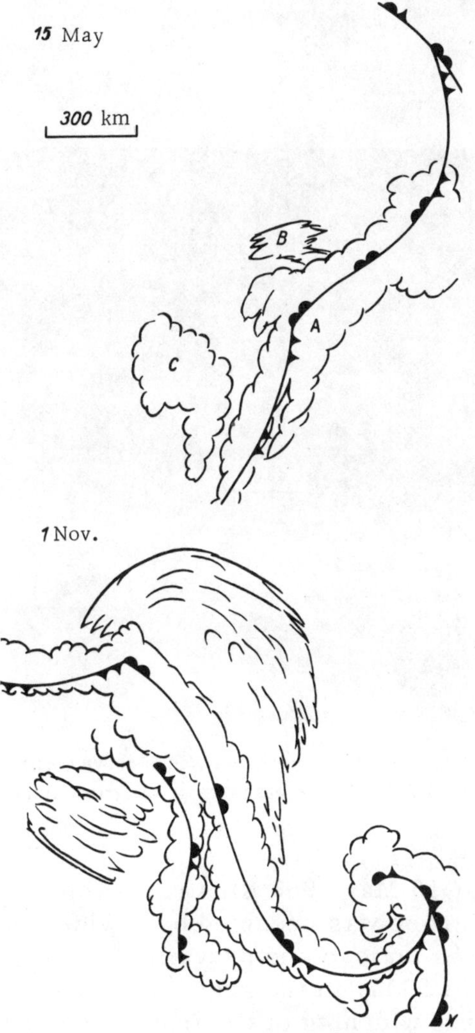

FIGURE 4. Kinds of cloud formation during cyclogenesis above the Danish straits on 15 May and 1 November 1969.

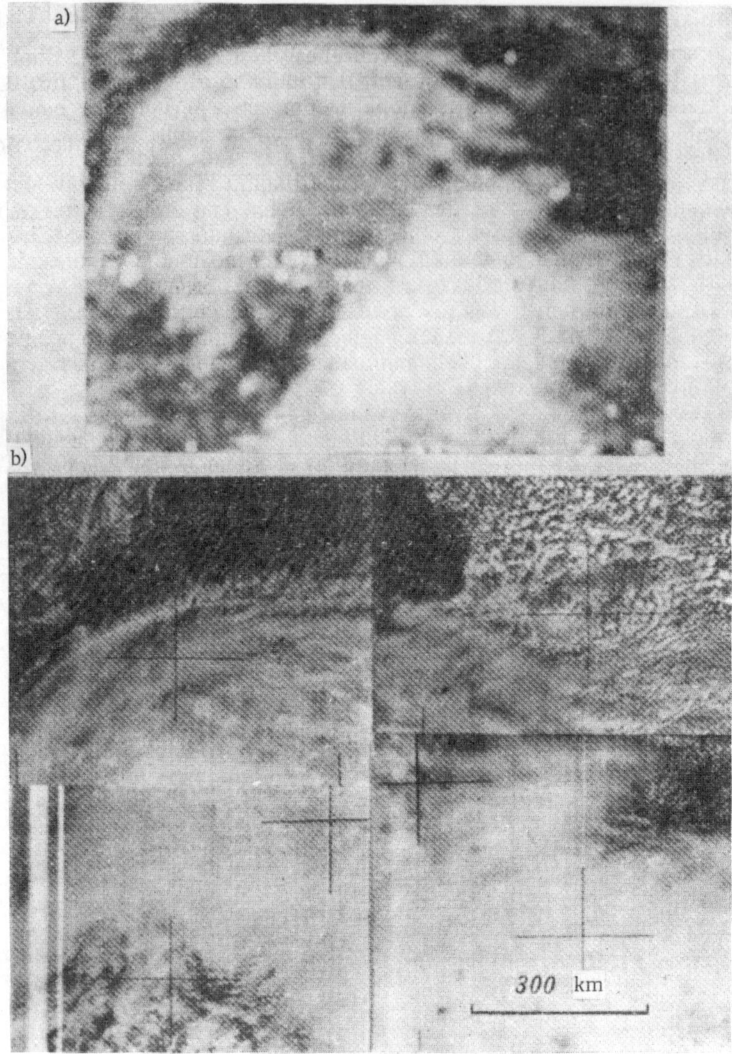

FIGURE 5. Cloud cap of a developing cyclone above Central Europe on
infrared (a) and TV (b) photographs taken by Meteor-8 at 1427 hrs on
27 April 1971.

The situation of 15 May 1969 illustrates the comma cloud formation. Northern Europe was occupied by a wide multicenter
depression, while an anticyclone prevailed above Central Europe.
The photograph of 15 May in the region of frontal zone A shows, in
addition to the usual widening of the frontal cloud band, a small cap
B made up of stratiformis and a cloud accumulation C of primarily
cumuliformis. The latter is slightly similar to a comma. It is
separated from the principal cloud band and plays the role here of

a secondary cloud vortex.   The presence of such a characteristic
pair of cloud formations (B and C) in the region of the wave is an
indication that cyclogenesis is under way.   Within 24 hours the wave
became a cyclone with spiral cloud pattern.   An example of TV
and infrared photographs of the cloud cap of a developing cyclone is
shown in Figure 5.

CONCLUSION

The cloud field shown by photographs obtained from artificial
satellites during cyclogenesis over Southeastern Europe and the
Danish straits has one common, very important feature.   The
inception of cyclones in both regions is accompanied by the
appearance of a cap stratiformis possessing a band structure with
anticyclonic curvature of the northern edge.   As a rule, the photograph shows a shadow band, which limits the cap from the north
and which points to a high upper limit of the clouds forming the cap.
Usually these are Ci, Cs and As.   Frequently the appearance of a
cap precedes the inception of cloud accumulations, primarily of
cumuliformis and usually in band form.

Preliminary analysis of temperature advection fields points to
a relationship between cloud caps with warm-air advection cells,
and to a relationship between the accumulation of cumuliformis
preceding these caps and cold air advection cells.

The authors wish to express their heartfelt thanks to engineer
T. D. Dzyubenko and senior laboratory assistants G. K. Kriskevich
and O. I. Mit'kina for their assistance in the data processing and
technical presentation of the work.

BIBLIOGRAPHY

1. Leonov, N.G. et al. Sinopticheskie usloviya sushchestvovaniya oblachnykh vikhrei (Synoptic
    Conditions for the Existence of Cloud Vortices).— Trudy Gidrometeorologicheskogo
    Tsentra SSSR, No.11. 1967.
2. Popova, T.P. and A.M. Tsar'kova. "Vtorichnyi" oblachnyi vikhr' (The "Secondary" Cloud
    Vortex).— Meteorologiya i Gidrologiya, No.9. 1967.
3. Popova, T.P. O strukture oblachnosti v tsiklonakh na yuge Evropeiskoi territorii SSSR po
    snimkam s IZS (Cloud Structure in Cyclones in Northern European USSR According to
    Satellite Photographs).— Trudy Gidrometeorologicheskogo Tsentra SSSR, No.36. 1968.

UDC 551.509

CLOUD EDDIES IN WAKES OF ISLANDS

T. Kh. Geokhlanyan

This paper is concerned with the formation of cloud eddies in wakes of single mountainous islands. Recommendations are given which yield useful results in forecasting the wind field from cloud patterns.

The advent of meteorological satellites, which supply pictures of clouds over large areas, has led to the discovery of orderly structures of cloud wakes produced by small islands (Figure 1). These cloud formations most frequently appear in the form of two parallel trails, within which the eddies are arranged in a staggered pattern and have mutually opposite circulations. The phenomenon of cloud eddies in the wakes of single mountainous islands is similar to Kármán vortex streets, observed under laboratory conditions when a cylinder is placed in an infinite air flow.

One of the latest experimental studies of the formation of vortex streets at different Reynolds numbers (Re) was carried out by Zdravkovich /14/, who studied vortex streets forming in a wind tunnel in the wake of a cylinder placed in a viscous fluid flow. The Reynolds number was slowly raised from 30 to 90. The flow regime was established by varying the fluid velocity. Figure 2 depicts an ideal vortex street obtained in /14/ at Re = 90.

As a result of many experiments the following mechanism was suggested for the inception of Kármán vortex streets: instability of [fluid] layers with shear at sufficiently high values results in the concentration of eddies at certain points. It was also discovered that a strong interaction exists in a developed vortex street between the eddies and the surrounding fluid. This interaction damps out with weakening of the actual eddies. At large distances from the body past which flow occurs the eddies elongate and become elliptical.

Vortex formations in the wake of a body undergo some evolution with increasing Reynolds number /4, 9, 13, 14/. Thus at Re = 30 the flow is laminar and there are no eddies. Starting with some critical value (Re $\simeq$ 50) a vortex pair appears directly downstream

of the obstacle in the direction of flow. The symmetric pattern is
replaced by one which is asymmetric, and vortices start in turn to
separate from the body and are then carried away by the flow. A
vortex street forms and persists at a large distance from the
obstacle, the distance between vortex rows increasing downstream.
When the Reynolds number becomes sufficiently high (Re = 100) no
eddies exist immediately at the body; they are replaced in this
location by a vortex-free space, which is then followed by the
standard vortex street. At even higher Re the eddies are rapidly
dissipated, so that they no longer exhibit the double-trail shape.
This "upper limit" of Re is not always the same: experimental
results in wind tunnels show that it does not exceed 2500 /9/. At
Re > 2500 individual vortices can no longer be distinguished and
the flow becomes fully turbulent, the systematic periodicity no
longer being observed.

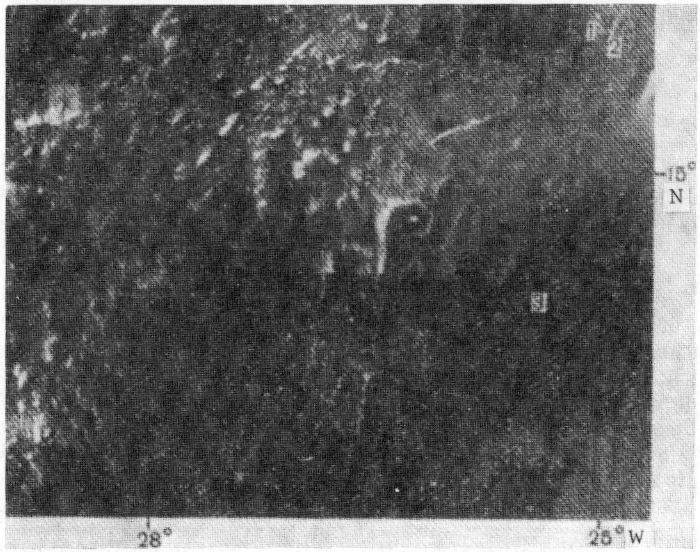

FIGURE 1. Eddies to the lee of Cape Verde Islands. Photograph from
the Cosmos-206 satellite on 31 March 1968 at 1351 hrs during orbit
253–254:

1 — Santo-Antão Island (1975 m); 2 — São Vicente Island (774 m);
3 — Fogo Island (2829 m).

Kármán vortex streets thus form over some range of Reynolds
numbers (35 < Re < 2500), their existence being determined by the
free-stream velocity u_0, kinematic viscosity in this stream v (in the

atmosphere ν is the eddy viscosity), as well as the characteristic dimensions of the obstacle d:

$$Re = \frac{u_0 d}{\nu}. \tag{1}$$

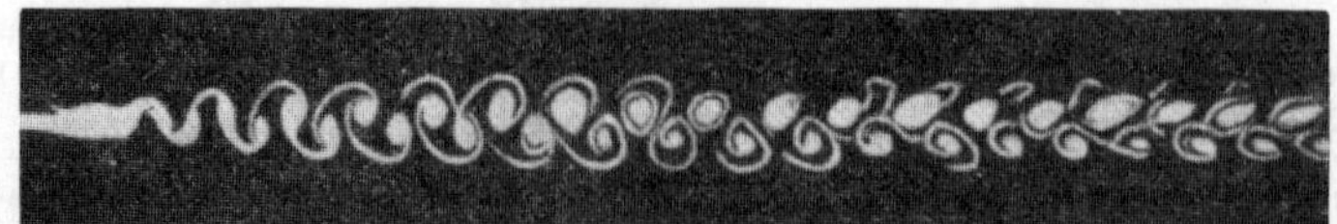

FIGURE 2. Formation of an ideal (Kármán) vortex street in the laboratory at Re = 90.

The same picture is also observed in the atmosphere: streets of vortex pairs form past single mountainous islands or chains of small islands and these are accompanied by clearings immediately past the islands. The dimensions of the clearings depend on the size and altitude of the island and on the variables determining the state of the atmosphere.

Questions as to the formation of vortex streets and the nature of clearings past individual islands and groups of islands of the Aleutian Islands are considered in /12/, where an attempt is made to establish the relationship between the mutual location and topography of the islands, the vertical stratification and the air-flow patterns. Noting that vortex streets generally form in satisfactory agreement with the two-dimensional theory, use is made in /12/ of additional hypotheses for explaining various aspects of this phenomenon. Thus the clearing observed immediately past the island is ascribed to the foehn effect, which is common in flow past mountains.

Study of a large volume of satellite cloud-cover information showed that clouds in vortex-street formations are most frequently encountered in subtropical latitudes, where old anticyclones with developed inversions are located. These vortex formations are observed over water and consist of stratocumulus. Clouds in vortex-street formation appear under conditions of weak vertical exchange and comparatively low wind speeds in the atmosphere in the vicinity of their formation. The width of this cloud vortex street depends on the effective* (relative to the wind) diameter of the island.

The first theoretical model of a vortex street was due to von Kármán /1, 3/, who solved the problem of the motion of a body (infinite cylinder) in an ideal fluid with the formation of eddies in the

* The effective diameter d_{ef} is the island diameter at an altitude of 30 m /8/.

body wake. Vortex streets are described by the following variables introduced by von Kármán (Figure 3): a is the distance between two adjoining vortices of the same trail forming one side of the vortex street; h is the distance between the trails; l is the distance between the corresponding vortices of both trails; $l^2 = \left[h^2 + \left(\frac{a}{2} \right)^2 \right]$; Γ is the strength of one vortex; N is the frequency of separation of vortex pairs; $N = \frac{u_l}{a} = \frac{1}{T}$; T is the time between separation of vortices; u_0 is the free-stream velocity; u_l is the velocity at which the vortices move downstream in a coordinate system whose origin is at the obstacle; and L is the length of the wake (defined as a limited region within which the disturbance is observed).

In spite of the fact that the Kármán theory disregards a number of important factors, primarily viscosity, estimates following from this theory give in a first approximation satisfactory agreement with experimental results and can be used in a number of cases for approximate calculations /6, 7, 10/.

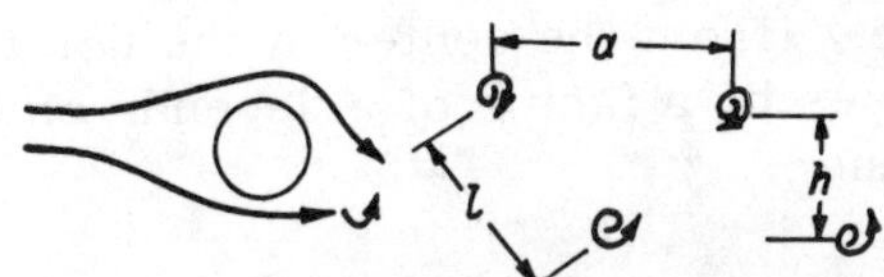

FIGURE 3. Schematic of the arrangement of vortices in flow about a cylinder.

On the other hand, if one considers the fluid viscosity, and hence eddy diffusion, it becomes possible to estimate the evolution of vortices as they move away from the body and the lifetime of an individual eddy. As is known from hydromechanics /2/, vorticity is described by an equation having the following form in cylindrical coordinates:

$$\zeta = \frac{1}{r} \frac{\partial (rv)}{\partial r}. \tag{2}$$

Suppose initially we have a velocity distribution corresponding to a straight vortex filament with vertical axis z and vortex strength Γ. Then in any subsequent period of time during this motion the projections of velocities on coordinate axes r, θ, z vary as follows: v_r and v_z are always equal to zero, while v_θ is a function of only r and t.

Under these conditions the equation defining the time variation of the vortex assumes the simplified form

$$\frac{\partial \zeta}{\partial t} = \nu \Delta \zeta = \frac{\nu}{r} \frac{\partial \left(r \frac{\partial \zeta}{\partial r} \right)}{\partial r},$$

(3)

where Δ is the two-dimensional Laplacian operator, while ν is the kinematic viscosity.

Equation (3) has a solution

$$\zeta = \zeta_0 e^{-r^2/4\nu t}$$

or

$$\zeta = \frac{\Gamma}{4\pi \nu t} e^{-r^2/4\nu t},$$

(4)

i.e., the eddy diffusion is a function of fluid viscosity and increases with time.

Here the distance r (from the center of the point vortex) at which the vorticity decreases by a factor of e depends on these quantities in the following manner:

$$r = 2\sqrt{\nu t}.$$

(5)

Assuming that a vortex of radius $r = \frac{h}{2}$ is still sufficiently perceptible, and using equation (5) we can estimate the eddy lifetime t:

$$t = \frac{h^2}{16\nu}.$$

(6)

If ν is expressed in terms of the coefficient introduced in /11/, namely

$$\beta = \frac{\nu N}{u_0},$$

(7)

then

$$\nu = \frac{\beta u_0}{N},$$

(8)

where β is a dimensionless coefficient not dependent on the dimensions of the obstacle and is equal to the ratio $\dfrac{S}{Re}$, where S is the Strouhal number $\left(S=\dfrac{Nd}{u_0}\right)$, while Re is the Reynolds number $\left(Re=\dfrac{u_0 d}{v}\right)$.

The numerator and denominator of (8) are now multiplied and divided by $N=\dfrac{u_l}{a}$; hence

$$v=\beta\left(\frac{u_0}{u_l}\right)^2 a^2 N. \tag{9}$$

Substitution of expression (9) into (6) yields

$$t=\frac{1}{16\beta N}\left(\frac{h}{a}\right)^2\left(\frac{u_l}{u_0}\right)^2. \tag{10}$$

The lifetime t_n of a given eddy can be expressed as follows:

$$t_n=\frac{D_n}{u_l}, \tag{11}$$

where D_n is the downstream distance from the center of the first to that of the given vortex.

Satellite photographs enable the values of a, h and D_n to be measured. Knowing also the values of v and β, equations (6)–(11) can be employed for calculating the free-stream velocity and the eddy lifetimes.

Cases of cloud vortex-street formation were singled out by examining a large volume of data obtained by USSR and USA satellites from 1968 to 1970. We detected 15 such cases which were most frequent in the wake of single mountainous islands.

Table 1 lists islands past which clouds with vortex-street formation were observed. The table lists the effective diameters (d_{ef}) of these islands, their maximum altitude above sea level (H), names of synoptic stations located on the islands and in their immediate vicinity, the indexes and geographic coordinates of the stations. Analysis of the relationships between H and d_{ef} for each island past which vortex streets were observed showed that this phenomenon occurred when $0.14 \leqslant \dfrac{H}{d_{ef}} \leqslant 0.04$.

TABLE 1. Geographic data for a number of islands past which vortex streets of clouds form

Island	d_{ef}, km	Maximum alti-tude of island (km)	Synoptic station			Index of syn-optic station	
			name	latitude	longitude	region	station number
Cheju	40	1.95	Cheju do	33°31'N	126°29'E	47	182
			Cheju do	33 31	126 32		184R
			Mosulpo	33 12	126 13		187
Yaku-shima	16	1.94	Yaku-shima	30 27	130 30		836
			Tanega-shima	30 44	131 00		837R
Madeira	40	1.86	Funchal	32 28	16 54	08	521R
			Porto Santo	33 03	16 20		524R
Cape Verde Islands:							
Santo-Antão	20	1.98	Mindêlo	16 53	25°00'W	08	583
Fogo	20	2.83					
Brava	8	0.97					
Santiago	16	1.39	Praia	14 55	23 31	08	589R
São Nicolau	14	1.30					
Sal	10	0.40	Sal	16 44	22 57	08	594R
Canary Islands:							
Gran Canaria	36	1.95	Las Palmas	27 56	15 23	60	030
Tenerife	40	3.70	Tenerife	28 29	16 20		015R
			Santa Cruz de Tenerife	28 27	16 15		020
Fuerteventura	20	0.78	Fuerteventura (pos Estanka)	28 31	13 53		035R
La Palma	20	2.36					
Guadeloupe	15	1.46	Ste. Rose	16 16	61 31	78	897R
West Spitsbergen (South Cape)	12	1.43					

N o t e. The letter R denotes radiosonde stations.

In these studies, no cases were discovered when the trails
forming the two sides of the vortex street mutually interfered and
washed each other away, although individual vortices are not
always clearly visible. This agrees with laboratory experiments
which showed that obstacles separated from one another at least
over a distance equal to one diameter of the obstacle-generated
isolated vortex trails, although these are close to one another. If,
however, one obstacle is situated upstream from the other, the
distance apart equaling several diameters, then both obstacles form
a single vortex street.

Sometimes, the vortices on the photograph appear to lie in an almost straight line. However, such an arrangement (as was shown in /15/) is in very precise agreement with synoptic data and the magnitude of the shift in time of the wind vector.

It was found by measuring the wake length on satellite photographs that the vortices were situated on the average 600–800 km from the island. Here the maximum (r_{max}) and minimum (r_{min}) vortex radii are respectively equal to 18 and 11 km.

When vortex street formations were detected on cloud photographs, we studied photographs taken over the given geographic region for the preceding and subsequent observation times. However, not a single case was observed when it was possible to follow the development of cloud vortex streets from the time of their inception to their complete disappearance. This regrettable circumstance only points to the fact that cloud formations do not exist naturally for more than 24 hours and that their development can be followed only by increasing the volume of satellite information. Here it was found that the calculated lifetime (t_n) of the farthest downstream eddy does not exceed 24 hours; on the average it is 20 hours, which is in agreement with the nature of the phenomenon.

It was noted above that the values of a, h, and D_n can be measured using an ordinary scale and having reference to the scale of the photograph. If the experimental values $\beta = 10^{-3}$ and $\nu = 10^3 \, \mathrm{m}^2/\mathrm{sec}$ are used together with measured values of a, h and D_n, equations (6), (10) and (11) can be employed to calculate free-stream velocity u_0, velocity of vortices within the street u_l, lifetime t_n of eddies, and time T between separation of successive vortices.

The calculations were carried out as follows: t_n was determined from equation (6), u_l from equation (11) and u_0 from equation (10). In accordance with wind-tunnel experiments it was assumed that $u_l = \dfrac{3}{4} u_0$ (according to laboratory observations $\dfrac{u_l}{u_0} \simeq 0.85$ /5/; the atmospheric vortex model yields $\dfrac{u_l}{u_0} = 0.75$, which lies between 0.71 (theoretical data) and 0.85 (laboratory data)), and since the frequency of vortex shedding is $N = \dfrac{u_l}{\tilde{a}} = \dfrac{1}{T}$, the time between vortex separations is $T = \dfrac{4\tilde{a}}{3u_0}$ (where $\tilde{a}$ is the average value of a). This expression was used when determining T.

All the above quantities were calculated for several cases of cloud vortex street formation in the wakes of Cheju do and Yaku-shima islands, located respectively in the northern and northeastern parts of the East China Sea.   The vortices in the wakes of these islands are here studied for the first time, while those associated with the Canary Islands and with the Island of Madeira were observed and studied on numerous previous occasions.   The results of calculations are given in Table 2.   To compare calculated and actual values of u_0 for windward sides of the islands, the latter were determined in each case from synoptic maps.

TABLE 2. Measured and calculated parameters. 1969 ($\tilde{a}$ and $\tilde{h}$ are average values)

Date	$\tilde{a}$, km	$\tilde{h}$, km	D_n, km	t_n	
				secs	hrs
Chaju do Island					
5 March	131	36	396	$8.1 \cdot 10^4$	22.5
17 March	77.4	24.4	288	$6.3 \cdot 10^4$	17.5
22 March	81	36	432	$8.1 \cdot 10^4$	22.5
Yaku-shima Island					
22 March	90	36	270	$8.1 \cdot 10^4$	22.5

u_l, m/sec	u_0 m/sec		L, km	T	
	predicted	actual		sec	hrs
Chaju do Island					
5	6.3	5	620	$1.3 \cdot 10^4$	3.6
8	10.2	10	504	$1 \cdot 10^4$	2.2
5	8	7	540	$2 \cdot 10^4$	5.5
Yaku-shima Island					
3.3	6	7	610	$2 \cdot 10^4$	5.5

Estimation of the free-stream velocity, which is a very important variable, is of great practical importance to the forecaster, who must be able to characterize the wind speed and direction, as well as the synoptic situation in a given region, from the external appearance of cloudiness, resorting to only a few geometric measurements.

On the basis of the above, when satellite photographs show the formation of cloud vortex streets in wakes of islands, the following must be taken into account: 1) this phenomenon occurs in the region of divergence of isobars near the ground, when the high and low pressure centers are located respectively to the left and right of the region under study; 2) cloud vortex streets form in the subinversion layer when the vertical exchange between layers is not too strong and consists of flow clouds (stratocumulus), situated from 0.5 to 2.0 km above the ocean surface; 3) the calculated free-stream velocities u_0 are close to the actual values observed in nature and amount to $7-10\,\text{m}/\text{sec}$; 4) the wake length L is on the average $600-800\,\text{km}$; here the direction of the cloud wake characterizes satisfactorily the air-flow direction in this region; 5) the 20-hour lifetime of the vortex street is quite sufficient for recording precisely the time of formation of new vortex pairs $\left(T=\dfrac{1}{N}\right)$ and for refining the remaining variables, in particular the ratio $\dfrac{u_l}{u_0}$, provided that more satellite data are made available.

BIBLIOGRAPHY

1. Kochin, N.E., I.A.Kibel', and N.V.Roze. Teoreticheskaya gidromekhanika (Theoretical Hydromechanics), Part 1. Moskva, GIFML. 1963.
2. Kochin, N.E., I.A.Kibel', and N.V.Roze. Teoreticheskaya gidromekhanika (Theoretical Fluid Mechanics), Part 2. Moskva, GIFML. 1963.
3. Lamb, H. Hydrodynamics. New York, Dover. 1945.
4. Bénard, H. Formation de centres de giration à l'arrière d'un obstacle en mouvement. — Comp. Rend., Vol.147, pp.839—842. 1908.
5. Birkhoff, G., and E.H.Zarantonello. Jets, Wakes and Cavities. New York, Academic Press. 1957.
6. Bowley, C.J., A.H.Glaser, R.J.Newcombe, and R.Wexler. Satellite Observations of Wake Formation Beneath an Inversion.— J. Atm. Sci., Vol.19, pp.52—55. 1962.
7. Chopra, K.P. and L.F.Hubert. Kármán Vortex-Streets in Earth's Atmosphere.— Nature, Vol.203, No.4952, pp.1341—1343. 1964.
8. Chopra, K.P. and L.F.Hubert. Mesoscale Eddies in Wake of Islands.— J. Atmos. Sci., Vol.22, pp.652—657. 1965.
9. Goldstein, S.(editor). Modern Developments in Fluid Dynamics, Vol.2. Oxford, Clarendon. 1957.
10. Hubert, L.F. and A.F.Krueger. Satellite Picture of Mesoscale Eddies.— Month. Weather Rev., Vol.90, p.457. 1962.
11. Lin, C.C. On Periodically Oscillating Wakes in the Oseen Approximation.— In: Studies in Fluid Mechanics Presented to R. von Mises, New York Academic Press, pp.170—176. 1959.
12. Lyons, W.A. and T.Fujita. Mesoscale Motions in Oceanic Stratus as Revealed by Satellite Data.— Month. Weather Rev., Vol.96, pp.304—314. 1968.
13. Nayler, J.L. and R.A.Frazer. Vortex Motion.— Advisory Committee for Aeronautics, R and M (New Series), No.332. 1917.
14. Zdravkovich, M.M. Smoke Observations of the Formation of a Kármán Vortex Street.— J. Fluid Mech., Vol.37, pp.491—496. 1969.
15. Zimmerman, L.I. Atmospheric Wake Phenomena near the Canary Islands.— J. Appl. Meteor., Vol.8, pp.896—907. 1969.

UDC 551.509

UTILIZATION OF SATELLITE INFORMATION IN SYNOPTIC PRACTICE

E. P. Dombkovskaya, V. F. Chernova

The case is considered of the inception and evolution of a mesoscale cloud vortex, leading to the development of nocturnal thunderstorms in a poorly delineated frontal zone at the periphery of an anticyclone.

It is now an accepted fact that TV and infrared cloud photographs must be used when analyzing synoptic processes in regions with sparse meteorological data. The question remains as to the utility of satellite data in regions with a dense network of meteorological stations. Can satellite data be used in synoptic practice and utilized in analysis and forecasting of weather?

The present authors have studied data on the utilization of satellite information in the operational forecasting practice of the Hydrometeorological Center of the USSR for August 1970, and believe that this question can be answered in the affirmative. The examined data show that, even in the presence of a dense network of meteorological stations (for example, in the USSR and Western Europe), satellite cloud data make it possible to refine and supply additional detail in the analysis of synoptic processes.

Published material and practice show that satellite cloud data aid in: a) refining the location of fronts on synoptic maps, particularly when the former are poorly expressed in temperature, wind, pressure tendency, etc. fields near the ground /2/; b) more clearly determining the stages in cyclone development /2−4/; c) obtaining an insight into the vertical cloud structure (in cases when both TV and infrared photographs are available) and into the nature of frontal interfaces /2, 3/; d) getting some idea as to the further development of synoptic processes /2, 4/; e) trying to understand the temperature stratification in the lower troposphere /2/, etc.

Analysis of data on the utilization of satellite cloud information in operational practice also showed that there are some other situations in which satellite cloud photographs are used in synoptic analysis. For example, cases were encountered when some details pertaining to the development of synoptic processes (which have a marked effect on weather) are easily detected on infrared photographs, but were not shown on synoptic maps. These include the inception of isolated cloud vortices of relatively small size (mesoscale vortices, "commas"), which are not directly related to cyclonic cells at the ground. It was shown in /4/ that these cloud vortices are generally located in the vicinity of intensive frontal zones and are most frequently observed behind a cold front within cold-air masses, while when combining with frontal cloud formations they induce a sharpening of the frontal interface or the formation thereat of a wave disturbance, sometimes even of an independent cyclone.

We shall now describe two cases of the inception of mesoscale vortices, one of which arose in the vicinity of an intensive frontal zone and can be classified as a common mesoscale vortex, while the other, an exception, arose at the periphery of an anticyclone, in the vicinity of a diffuse cold front. When the vortex approached the frontal zone, it did not merge with the latter but continued to exist independently in the immediate vicinity of the front.

I. A mesoscale vortex originated in the vicinity of a diffuse cold front over northern European USSR on 4–6 August 1970. The cold front which swung past an anticyclone over Northern Europe from the east and south was followed with difficulty on the ground synoptic map as soon as 0300 hrs on 3 August, particularly its southern part (there was no convergence of air currents, the temperature contrast between air masses separated by this front amounted to 4–6° at the ground and 3–5° at the 850-mb level, no thundershowers or storms were observed at the front). It could only be followed on infrared photographs taken at night close to the time for which the synoptic map was made, and was tracked with the aid of a cloud band which was quite wide (about 200 km) and bright about eastern Archangel Region and Komi ASSR, and a narrow (about 50 km) band consisting of broken gray cumulus at the latitude of Leningrad.

During the night of 4 to 5 August a small isolated zone of thundershowers and storms appeared in the Vologda region behind the diffuse cold front. This zone moved during 4–6 August along the periphery of the anticyclone, first to the southwest and then to the west and northwest through the Kalinin and Leningrad regions to Karelian ASSR.

No visible causes could be detected for the appearance of this isolated zone of thundershowers and storms, or for its quite prolonged persistence on ground synoptic maps and on pressure maps of the lower troposphere.

These showers and storms could not be classified as frontal for the following reasons: 1) the front was diffuse, was followed on maps only with difficulty, and no conditions existed for its sharpening; 2) the storms and showers were localized in a small region; 3) no conditions were observed which would favor the inception of a frontal wave in this region; in addition, the direction in which the precipitation zone moved (from the west and northwest) contradicts the assumption of a wave disturbance.

There was no basis for classifying these showers and storms as air-mass phenomena because: 1) they were observed primarily at night, ceasing by day during the maximum development of convection (1500 hrs); 2) an extensive fog zone was observed at the periphery of the anticyclone in the immediate vicinity of the above storm cell to the south and southwest (see circular map for 5 August), pointing to the presence of stable atmospheric stratification in the ground layer during the night and evening hours. This fog zone persisted for the entire morning and is seen clearly on a TV photograph taken at 1100 hrs. At the same time infrared photographs suggest that the causes for this phenomenon should be sought in processes occurring in the upper troposphere.

As of 0300 hrs on 4 August, i. e., the time of appearance of showers and storms, the straight chain of cumulus at the southern periphery of the anticyclone acquired the shape of a small vortex of radius about 200 km, located above the shower and storm cell. While the clouds were acquiring their spiral shape the image became much brighter, which points to increased cloud top heights.

At the ground and up to altitudes of 3 km no cyclonic circulation or perceptible trough was observed in these regions (Figure 1).

Analysis of pressure maps clearly showed that the formation of the cloud vortex can be attributed to the inception of a small, closed region of cyclonic circulation and an isolated cold cell above these regions, first in the lower and then in the upper troposphere (at the 200, 300 and 500 mb levels) against a background of high geopotential values.

During the following two days this closed region of cyclonic circulation was observed to move along the periphery of the anticyclone and extend to its lower levels: at 1500 hrs on 4 August it was detected at the 700-mb level, while at 0300 hrs on 5 August it was also located at the 850-mb level. The area occupied by closed cyclonic circulation at the 850-mb level did not exceed 200 km in diameter and was very shortlived (at 0300 hrs on 6 August it could no longer be detected on the 850-mb chart).

The mesoscale cloud vortex and its associated shower and storm cell moved with the velocity of winds at the 850-mb and surface

layer. Here the infrared photographs showed that during the first
36 hours the shower and storm cell on the ground synoptic map, the
closed region of cyclonic circulation at high altitudes, and the mesoscale cloud vortex moved simultaneously, while as of 0300 hrs on
6 August the shower and storm cell and its associated cloud vortex
slowed down somewhat and were found to be at the rear of the
cyclonic circulation region in straight air flows; as of 1500 hrs on
6 August they disappeared.

It was mentioned previously that storm activity and showers
associated with the mesoscale vortex ceased during daytime, which
is the time of maximum development of convection. Objectively
this is attributable to restructuring of temperature stratification
from night to day conditions, induced by the presence of cloudiness
and intensive radiation from its upper boundary at night /1/. The
sharpest differences in the vertical temperature distribution were
observed in the lower, 3-kilometer layer of the troposphere.

TABLE 1. Temperature lapse rate (deg/100 m) between principal isobaric surfaces

Date	Time	Radiosonde location	Level (mb)			
			Ground–850	850–700	700–500	500–300
4 August	0300 hrs	Vologda	0.33	0.67	0.70	0.70
	1500 hrs	Bologoe	0.87	0.33	0.95	0.67
5 August	0300 hrs	Bologoe	0.27	0.73	0.80	0.70
	1500 hrs	Leningrad	0.73	0.53	1.05	0.55
6 August	0300 hrs	Petrozavodsk	0.33	0.67	0.75	0.70

Table 1 lists average vertical temperature lapse rates between
the principal isobaric surfaces in the region above which the mesoscale vortex was situated. It is seen from the table that at night
(when showers and storms occurred) the temperature lapse rate
increased sharply from the 1000–850 mb to the 850–700 mb layer,
remaining unchanged above it, while by day the lapse rate in the
850–700 mb layer was smaller than at the ground in the upper half
of the troposphere and produced a quite thick barrier layer, which
prevented the development of thermal convection. These data on
the vertical temperature distribution on days with nighttime storms
are in agreement with observational data collected in Novosibirsk
/5/.

Yet another characteristic feature of the evolution of the mesoscale cloud vortex under study is noteworthy. The movement of
the vortex from the Vologda to the Leningrad Region and Karelian
ASSR during this period of time was accompanied by the arrival

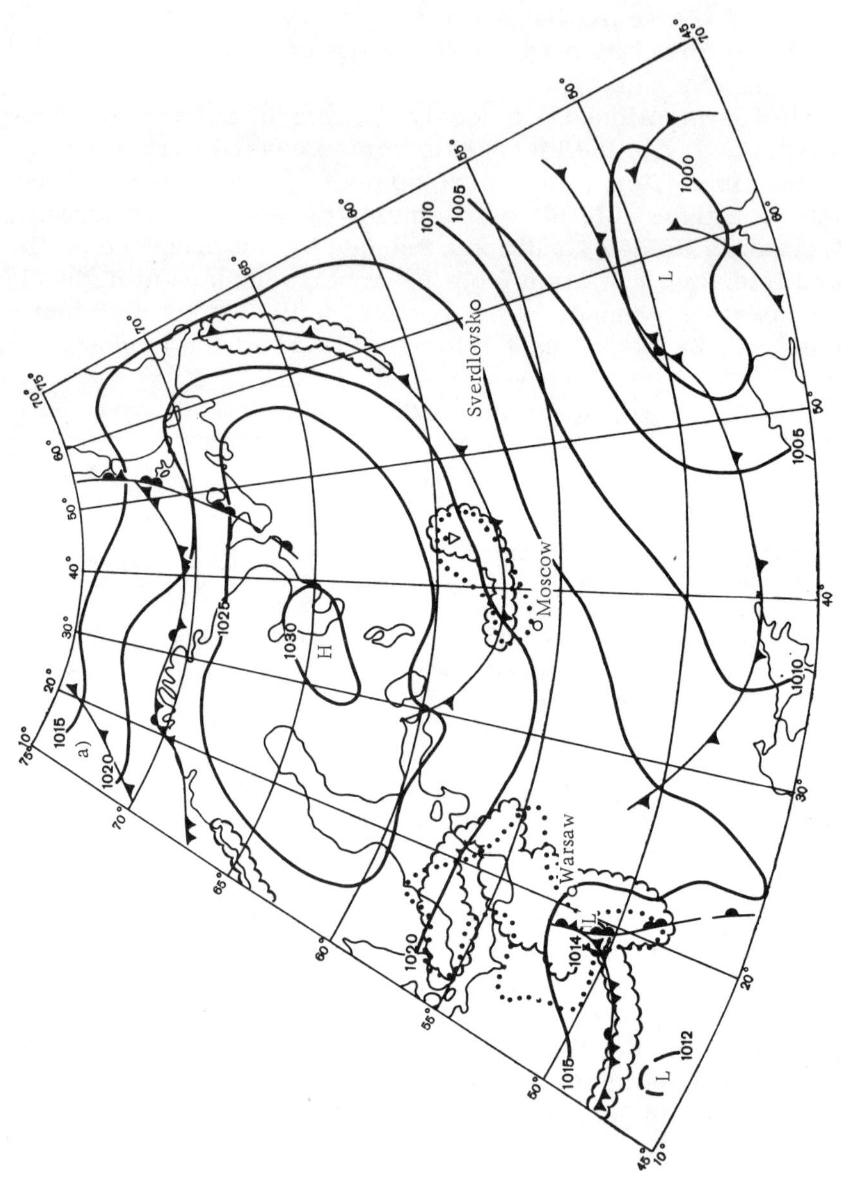
Sverdlovsko
Moscow
Warsaw
H
L
1000
1005
1010
1012
1014
1015
1020
1025
1030
a)

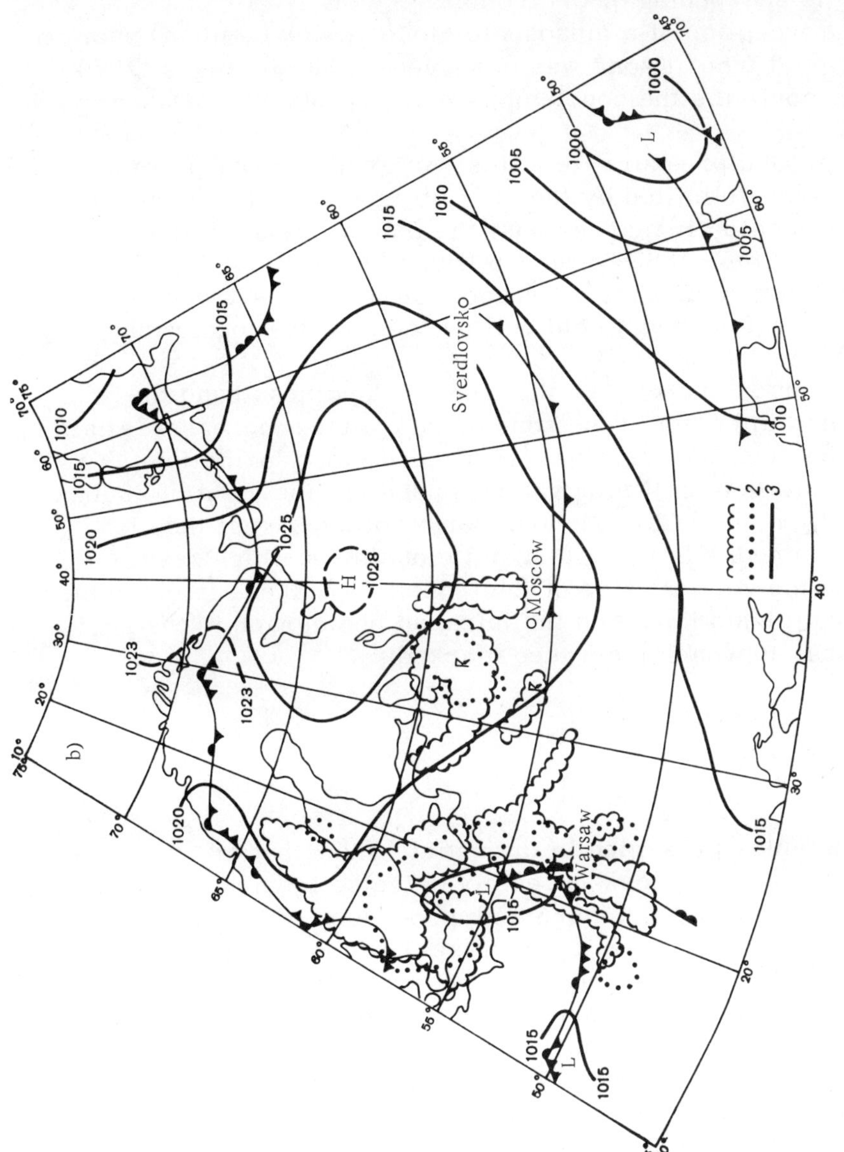

FIGURE 1. Synoptic maps and nephanalysis:

a — 0300 hrs, 4 August 1970; b — 0300 hrs, 5 August 1970; 1 — cloud mass boundaries; 2 — precipitation zone boundaries (from data of 12-hour totals); 3 — isobars.

over these regions of another cloud formation — a cloud system associated with a warm front and an occluded front and located on 4 August over Poland (Figure 1a). When these cloud formations approached one another (Figure 1b) they did not merge and there was also no sharpening of the frontal system.

II. The inception of a mesoscale cloud vortex (comma) near an intensive cold frontal zone was observed on 28–30 August 1970. This case confirms the conclusions made by others /4/ about such vortices.

In the ground pressure field this vortex has corresponding to it a small cyclone, delimited by the 995-mb isobar. This vortex was generated in cold air masses past the frontal cloud band which passed from Komi ASSR to the Ob River drainage basin. On pressure maps this vortex has corresponding to it closed cyclonic circulation regions and a cold cell on the relative topography map of 1000/500 mb.

As the vortex approaches the frontal cloudiness band the precipitation intensity in the zone of this vortex increases, attaining a maximum upon the combination of these cloud formations. A total precipitation of 46 mm was recorded for the night of 29 August.

Toward the end of the 24-hour period of 29 August the cloud vortex combined with the cold-front cloudiness, which resulted in increasing the vertical size of the frontal cloudiness (the brightness of the frontal band image on the infrared photograph in the region of their mutual approach increased markedly).

CONCLUSION

The examples presented in the article show that macroscopic and synoptic-scale processes, easily detected by means of ordinary synoptic data, coexist in nature with processes of smaller scales (meso- or subsynoptic-scale processes), which have a marked effect on the weather but cannot be discerned on synoptic maps even in regions with a very dense network of meteorological stations.

A great help in this respect is provided by satellite cloudiness pictures. These pictures provide continuous cloudiness data and make it possible to detect any cloud formations the dimensions of which are larger than the resolving power of the equipment.

BIBLIOGRAPHY

1. Vel'tishchev, N.F. Yacheikovaya konvektsiya v atmosfere (Cellular Convection in the
     Atmosphere).— Trudy Gidrometeorologicheskogo Tsentra SSSR, No.50. 1969.
2. Vel'tishchev, N.F. and T.P.Popova. Ispol'zovanie dannykh nablyudenii ob oblachnosti
     so sputnikov v sinopticheskom analize (Utilization of Satellite Cloud Data in Synoptic
     Analysis).— Metodicheskoe pis'mo. Leningrad, Gidrometeoizdat. 1970.
3. Dombkovskaya, E.P. Interpretatsiya infrakrasnykh snimkov oblachnosti, poluchennykh so
     sputnika "Kosmos-122" (Interpretation of Cosmos-122 Infrared Cloud Pictures).—
     Meteorologiya i Gidrologiya, No.5. 1968.
4. Minina, L.S. Praktika nefanaliza (Nephanalysis Practice). Leningrad, Gidrometeoizdat.
     1970.
5. Yagudin, A.Ya. Fiziko-statisticheskoe issledovanie nochnykh groz (A Physicostatistical Study
     of Nocturnal Thunderstorms).— Meteorologiya i Gidrologiya, No.6. 1970.

UDC 551.509

SOME CHARACTERISTICS OF THE VERTICAL STRUCTURE OF THE HUMIDITY FIELD OVER THE NORTH ATLANTIC

E. P. Dombkovskaya

The article considers the average distribution of specific humidity over the western part of the North Atlantic (from data of the A, C, D and E weather ships) during the warm period; certain statistical characteristics of the specific humidity are presented. A comparison is made with the characteristics of the vertical structure of the humidity field over land.

The solution of inverse meteorological problems, in particular for reconstructing the vertical humidity distribution from satellite data, requires the availability of information on the spatial structure of the humidity field. Since the variations in the humidity profile are random, these data can be obtained only by examining the statistical characteristics of the humidity field. These are:
1) average specific humidity $\overline{q}$ at different atmospheric levels,

$$\overline{q}\,(p_k) = \left(\frac{1}{n}\right) \sum_{i=1}^{n} q_i\,(p_k);$$

2) their rms deviations,

$$\sigma_q\,(p_k) = \sqrt{B_{qq}\,(p_k,\ p_k)};$$

3) correlations between deviations of the mean at different levels, expressed in the form of autocorrelation matrices of the specific humidity,

$$B_{qq}\,(p_k,\ p_l) = \frac{1}{n} \sum_{i=1}^{n} q_i'\,(p_k)\,q_i'\,(p_l),$$

where $q_i\,(p)$ are values of the specific humidity, $q_i'\,(p)$ deviations of the specific humidity from normal, p the pressure, and n the

number of measurements of the vertical humidity distribution.
Subscripts k and l pertain to the different atmospheric levels.

In the analysis it is more convenient to use autocorrelation
matrices:

$$r_{qq}(p_k,\ p_l) = \frac{B_{qq}(p_k,\ p_l)}{\sigma_q(p_k)\ \sigma_q(p_l)}.$$

The statistical characteristics and the vertical structure of the
humidity field were studied by various authors /1, 2, 4–6/. The
majority of calculations were made for January and July, with no
subdivision into cloudy and clear days.

The calculations in /4/ were carried out using data of land
stations located in different climatic zones of the Soviet Union.
Reference /1/ presents results of calculations using data of
weather ships of the North Atlantic and land stations in Western
Europe; in this study the correlation matrices were first
calculated separately for cases with clear and cloudy skies (for
January and July). The results of calculations show that the
correlation coefficients r_{qq} for cases with cloudy and clear weather
differ, and that this difference increases somewhat with altitude.

In all these studies attention is called to the substantial
variability of r_{qq} and other statistical humidity characteristics as a
function of the specific conditions for which these were calculated
(latitude, conditions of atmospheric circulation, season, etc.). The
conclusion was also reached that the use of correlation functions in
reconstructing the vertical moisture distribution when the former
do not correspond to specific conditions can result in substantial
errors. It hence appears desirable to calculate the statistical
characteristics of the vertical structure of the humidity field for
different seasons, latitude, and weather conditions.

During the past few years the feasibility has been discussed in
scientific literature of utilizing satellite observations of microwave
radiation for reconstructing vertical humidity and temperate lapse
rates. Since at present the radiometer data used in meteorology
are exclusively those obtained over the sea surface, the greatest
interest with respect to this problem is presented by a priori
information on the vertical structure of meteorological elements
above oceans.

This article presents statistical characteristics of the specific
humidity ($\bar{q}$, σ_q, B_{qq}, r_{qq}) calculated from radiosonde data of weather
ships over the North Atlantic. The ships whose data were used are:
A (62°N, 33°W), C (52°N, 35°W), D (44°N, 41°W) and E (35°N, 48°W)
for the warm season (May and June, which correspond to the end of

spring and start of summer; July and August, which represent summer; September and October corresponding to autumn) under clear skies.

Since the frequency of days with clear skies over the North Atlantic is low (particularly to the north of 50°N) /3/, the number of vertical humidity profiles taken by ships A and C was found to be insufficient for statistical processing. However, since the differences in the physicogeographic conditions and conditions of atmospheric circulation in the regions covered by ships A and C are insignificant, it was possible to combine these data into a single statistical series. A total of 220 cases were selected for the three years 1965–1967. The cases chosen for statistical processing were 56 from ships A and C, 70 from ship D, and 94 from ship E.

In order to clarify the extent to which r_{qq} is affected by latitude alone, the correlation matrices were calculated separately for each of the singled-out regions (region of ships A and C, ship D and ship E) for all the six months; to determine the time variation in r_{qq} the correlation matrices were calculated from data of all four ships separately for each season.

HUMIDITY DISTRIBUTION IN THE LOWER HALF OF THE TROPOSPHERE OVER WESTERN REGIONS OF THE NORTH ATLANTIC

The decrease in the average specific humidity with altitude, characteristic of the troposphere above land /4, 5/, also occurs over oceans. A distinguishing sign of the vertical variation in specific humidity over the ocean is a more rapid decrease in humidity with altitude as compared with land, particularly perceptible in the lower 1.5 km layer (Figure 1). The sharp decrease in specific humidity in the 700–500 mb layer $(\bar{q}_{700}/\bar{q}_{500} \approx 3)$, observed in /4, 5/ and attributed there to the coincidence of the mean height of cloud tops with the 500 mb surface, is observed also in the cases with clear weather considered here (Table 1).

Moreover, in clear weather ratios $\bar{q}_{700}/\bar{q}_{500}$ even exceed the value obtained in /4, 5/ as the average for different weather conditions. Apparently, the causes for the higher values of $\bar{q}_{700}/\bar{q}_{500}$ should be sought elsewhere. In this author's opinion it may be due to the different thickness of layers for which these quantities are calculated: the 1000-850 and 850–700 mb layers are approximately 1.5–1.8 km thick, while the thickness of the 700–500 mb layer is approximately 2.5 km. This explanation is also confirmed by the values of the vertical gradient of specific humidity calculated from data of /4/: the vertical specific humidity gradient decreases with altitude.

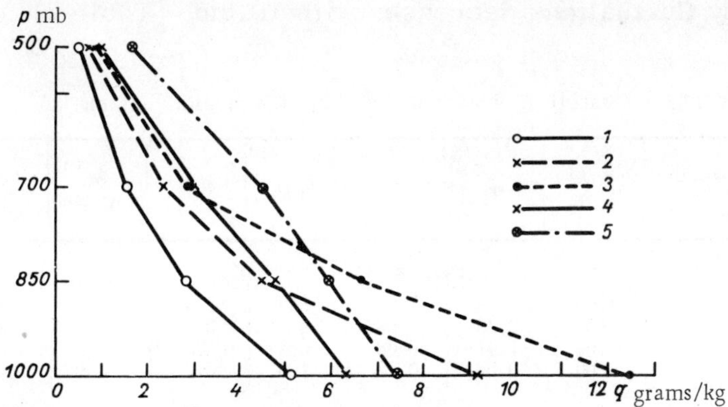

FIGURE 1. Average vertical profile of water vapor (q, grams/kg):

1 — weather ship A; 2 — weather ship D; 3 — weather ship E;
4 — Murmansk; 5 — Alma-Ata.

A second characteristic feature of the vertical humidity distribution over oceans as compared with that over land (also most clearly perceptible in the lower 1.5 km layer) is the very great dependence of the average specific humidity at different altitudes on latitude and, consequently, on surface temperature. The average value $\bar{q}$ of the specific humidity over oceans in the lower half of the troposphere increases perceptibly from north to south, which is not true of the specific humidity over land (Figure 1).

TABLE 1. Ratios $\bar{q}_n/\bar{q}_{n+1}$ for the principal isobaric surfaces over land and sea (values of $\bar{q}_n/\bar{q}_{n+1}$ for land were calculated from data of /4/)

Layer (mb)	$\bar{q}_n/\bar{q}_{n+1}$	
	land	sea
1000—850	1.27—1.45	1.82—2.04
850—700	1.35—1.59	1.82—2.55
700—500	2.38—2.97	2.75—3.30

The variability in the specific humidity can be characterized by the amplitude of fluctuations and by the variance. The magnitude of fluctuations of specific humidity was shown to increase together with an increase in the specific humidity proper. Here, unlike over land where maximum specific humidity fluctuations are observed in the 700–500 mb layer /4/, the most appreciable fluctuations over oceans occur at the former's surface. The amplitude of specific humidity fluctuations decreases with altitude (Table 2).

TABLE 2. Statistical characteristics of specific humidity (grams/kg)

Weather ship	Level (mb)	$\bar{q}$	σ_q	q_{max}	q_{min}	Δq (max − min)	$\dfrac{\sigma_q}{\bar{q}} \cdot 100\%$
				Number of cases, 56			
A, C	1000	5.28	1.03	8.3	2.4	4.9	19.6
	850	2.89	0.88	5.8	0.7	5.1	30.4
	700	1.41	0.68	3.4	0.3	3.1	48.0
	500	0.50	0.23	1.5	0.2	1.3	46.0
				Number of cases, 70			
D	1000	9.12	2.71	15.0	4.5	10.5	29.5
	850	4.57	2.11	9.5	1.0	8.5	46.2
	700	2.28	1.32	5.8	0.6	5.2	57.8
	500	0.74	0.47	2.7	0.2	2.5	63.4
				Number of cases, 94			
E	1000	12.42	2.31	16.6	5.3	11.3	18.6
	850	6.66	2.17	12.0	1.2	10.8	33.3
	700	2.78	1.62	7.0	0.6	6.4	58.2
	500	0.84	0.49	3.2	0.2	3.0	58.4

TABLE 3. Average values of specific humidity in the western part of the North Atlantic and its time variability

Level (mb)	$\bar{q}$	σ_q	q_{min}	q_{max}	Δq (max − min)	$\dfrac{\sigma_q}{\bar{q}} \cdot 100\%$
			May−June			
1000	8.72	3.01	2.4	14.8	12.4	34.5
850	4.30	2.31	0.9	9.5	8.5	53.6
700	1.99	1.43	0.4	6.5	6.1	71.6
500	0.61	0.34	0.2	1.8	1,6	55.7
			July−August			
1000	9.59	3.45	3.9	16.6	12.7	35.9
850	5.16	2.42	1.6	10.8	9.2	46.9
700	2.35	1.30	0.6	6.4	5.8	55.2
500	0.72	0.40	0.2	3.2	3.0	55.7
			September−October			
1000	8.64	4.06	2.7	15.9	13.2	46.8
850	4.28	2.37	0.7	9.8	9.1	55.2
700	1.94	1.47	0.3	7.0	6.7	75.6
500	0.70	0.52	0.1	2.7	2.6	74.2

The variance of specific humidity (as shown by data of Table 2) also decreases with altitude, while the relative variability $\sigma_q/\bar{q}$ (coefficient of variance) increases with altitude, attaining a maximum ($44-63\%$) in the $700-500$ mb layer. The greatest relative variability in specific humidity is observed in the region of ship D where a rapid change of air masses, with markedly differing temperature and moisture content, occurs more frequently than in other regions of the North Atlantic.

Some idea about the time variation in the specific humidity can be obtained by examining Table 3. The time variation in the specific humidity from spring to autumn has, at all altitudes, the same shape as the summer maximum. The span of fluctuations in the specific humidity from spring to summer and autumn changes little. The relative variability of the specific humidity at all altitudes in autumn is greater than in summer and spring; here the minimal variability at the 850, 700 and 500 mb levels occurs in summer, while at the 1000 mb level it takes place in spring. The variances of the specific humidity in the ground layer increase from spring to autumn, while in all other layers they vary little.

CORRELATION BETWEEN DEVIATIONS FROM THE MEAN SPECIFIC HUMIDITY AT DIFFERENT ATMOSPHERIC LEVELS

The correlation between deviations from the mean specific humidity, $q'_i(p_k) = \bar{q}(p_k) - q_i(p_k)$, is described by normalized auto-correlation matrices of the specific humidity:

$$r_{qq} = \frac{B_{qq}(p_k,\ p_l)}{\sigma_q(p_k)\,\sigma_q(p_l)}.$$

Analysis of normalized autocorrelation matrices (Table 4) showed that the closest correlation between $q'_i(p_k)$ is observed for adjoining levels. As the distance between the levels being correlated becomes greater, r_{qq} decrease; here the latter, calculated for the entire period under study for individual latitude zones (Table 4a$-$c), decrease quite perceptibly with altitude (for northern regions r_{qq} ranged from 0.6 at the 850 mb level to 0.1 at the 500 mb level; for the southern regions this variation is from 0.7 to 0.2, respectively). On the time scale (Table 4d$-$f) a sharp reduction in r_{qq} with altitude is observed only in spring (from 0.7 to 0.2). In summer and particularly in autumn the correlation coefficients between the deviations of specific humidity from the average at the 1000 mb level and at remaining levels remain quite high (for example, at the 500 mb level in summer $r_{qq} = 0.32$, while in autumn $r_{qq} = 0.46$).   The correlation coefficients between q'_{1000} and $q'(p_k)$ at the remaining isobaric surfaces, calculated for different latitude zones of the North Atlantic, increase somewhat from north to south (Figure 2).

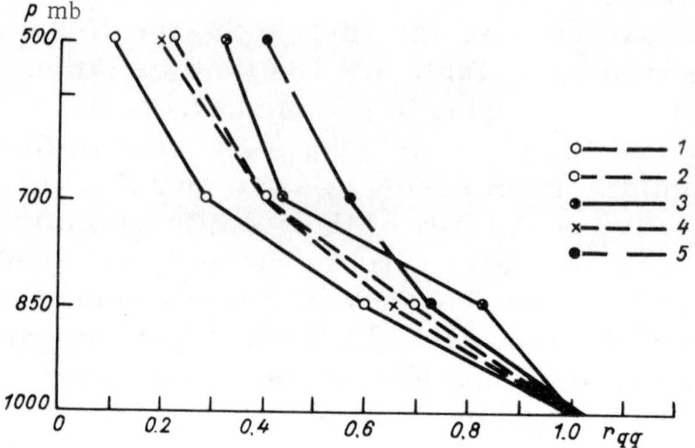

FIGURE 2. Vertical sections of the autocorrelation coefficients of specific humidity:

1 — weather ships A and C (May—October); 2 — ship E (May—October; 3 — weather ships A, C, D and E (July—August); 4 — Murmansk (July); 5 — Kzyl-Orda (July).

The values of r_{qq} for different seasons (over the entire western part of the North Atlantic) exceed the values of r_{qq} calculated from data of individual ships for the entire warm season, and increase from spring to autumn (Table 4a and b).

The above discussion implies that the autocorrelation coefficients of specific humidity depend more on season than on latitude.

It is seen by comparing vertical sections of the autocorrelation coefficients of specific humidity with the data of /4/ for land that, unlike average vertical humidity profiles, the variability of elements of correlation matrices as a function of latitude is more perceptible over land than over the ocean (Figure 2).

TABLE 4. Normalized autocorrelation matrices r_{qq} of specific humidity

p mb	1000	850	700	500
a) Ships A and C, 56 cases				
1000	1.00	0.60	0.29	0.10
850		1.00	0.34	0.06
700			1.00	0.52
500				1.00
b) Ship D, 70 cases				
1000	1.00	0.66	0.33	0.14
850		1.00	0.40	0.12
700			1.00	0.42
500				1.00
c) Ship E, 94 cases				
1000	1.00	0.70	0.40	0.22
850		1.00	0.58	0.36
700			1.00	0.56
500				1.00
d) May−June, 54 cases				
1000	1.00	0.78	0.45	0.22
850		1.00	0.54	0.33
700			1.00	0.53
500				1.00
e) July−August, 96 cases				
1000	1.00	0.83	0.44	0.32
850		1.00	0.53	0.33
700			1.00	0.44
500				1.00
f) September−October, 70 cases				
1000	1.00	0.82	0.53	0.46
850		1.00	0.58	0.30
700			1.00	0.68
500				1.00

BIBLIOGRAPHY

1. B o l d y r e v , V.G. and V.M.O l e s h e v. O statisticheskoi strukture vertikal'nykh profilei
temperatury i vlazhnosti (On the Statistical Structure of the Vertical Temperature and
Humidity Distributions).— Trudy MMTs, No.11. 1966.
2. B o l d y r e v , V G., L.I.K o p r o v a, and M.S.M a l k e v i c h. Ob uchete variatsii vertikal'nykh
profilei temperatury i vlazhnosti pri opredelenii temperatury podstilayushchei
poverkhnosti po ukhodyashchemu izlucheniyu (Consideration of Vertical Temperature

and Humidity Profiles in Determining the Surface Temperature from Departing Radiation).— Izvestiya AN SSSR, Fizika Atmosfery i Okeana, No.7. 1965.

3. L o b a n o v a , V.Ya. (editor). Klimaticheskie karty oblachnosti severnogo polushariya (Climatic Maps of Cloudiness for the Northern Hemisphere). Moskva, NIIAK. 1967.

4. K o m a r o v , V.S. Nekotorye statisticheskie kharakteristiki vertikal'nykh profilei temperatury i vlazhnosti (Some Statistical Characteristics of the Vertical Temperature and Humidity Profiles).— Trudy NIIAK, No.58. 1969.

5. K o m a r o v , V.S. Raspredelenie vodyanogo para v svobodnoi atmosfere (Water-Vapor Distribution in the Free Atmosphere).— Trudy NIIAK, No.47. 1969.

6. P o p o v , S.M. Nekotorye statisticheskie kharakteristiki vertikal'noi struktury polei temperatury i vlazhnosti (Some Statistical Characteristics of the Vertical Structure of Temperature and Humidity Fields).— Izvestiya AN SSSR, Fizika Atmosfery i Okeana, No.1. 1965.

UDC 551.509:551.521.32

EXPERIMENTS ON INCORPORATING RADIATIVE HEAT INFLUX IN NUMERICAL FORECASTING

P. N. Belov, L. V. Berkovich, R. L. Alpatova,
L. S. Bushueva

Data on the contribution of radiative and eddy heat fluxes to pressure variations are presented and analyzed; the material was obtained from experiments with a six-level forecasting model using general equations. Several versions are examined of making allowance for heat influx in three-day forecasts for the Northern Hemisphere.

Radiative heat influx is an important factor affecting long-range weather changes. The need for incorporating it in long-term weather forecasts is obvious. The different ways of making allowance for the radiative heat influx in weather forecasting and in simulating the general circulation of the atmosphere were considered in studies by Blinova /6, 7/, Kibel' /12/, Marchuk /14/, Mashkovich /15/ and others /9—11, 13, 16—26/.

The question of the role of radiative, and in general of non-adiabatic factors in short-term weather variations, as well as in variations in the weather over some average time period are not so clear. Some studies present data which point to a substantial effect of nonadiabatic factors in these cases /2, 17, 20/. However, this question requires special investigations to derive quantitative characteristics of the effect of nonadiabatic factors on variations in meteorological variables. Some characteristics of the effect of the radiative, and particularly also of eddy heat influx with respect to pressure changes at different atmospheric levels for three days are presented in this article. These characteristics were obtained by numerical experiments with a six-level nonadiabatic forecasting model using general equations. We start with a brief examination of a version of an adiabatic forecasting model, worked out by Berkovich.

SIX-LEVEL FORECASTING MODEL EMPLOYING GENERAL EQUATIONS

We use the following system of equations of atmospheric dynamics:

$$\frac{\partial u}{\partial t} - lv = -m\left(\frac{\partial u^2}{\partial x} + \frac{\partial uv}{\partial y} + \frac{\partial u\omega}{\partial p} + \frac{\partial \Phi}{\partial x}\right), \tag{1}$$

$$\frac{\partial v}{\partial t} + lu = -m\left(\frac{\partial uv}{\partial x} + \frac{\partial v^2}{\partial y} + \frac{\partial v\omega}{\partial p} + \frac{\partial \Phi}{\partial y}\right), \tag{2}$$

$$\frac{\partial u}{\partial x} + \frac{\partial v}{\partial y} + \frac{\partial \omega}{\partial p} = 0, \tag{3}$$

$$\frac{\partial \theta}{\partial t} = -m\left(\frac{\partial u\theta}{\partial x} + \frac{\partial v\theta}{\partial y} + \frac{\partial \omega\theta}{\partial p}\right) + \left(\frac{P}{p}\right)^{\lambda}\frac{\varepsilon}{c_p\rho}, \tag{4}$$

$$\frac{\partial \Phi}{\partial p} + \frac{R}{p}\left(\frac{p}{P}\right)^{\lambda}\theta = 0, \tag{5}$$

where u, v and ω are the wind-speed components along the x, y and p axes; t is time; $l = 2\Omega \sin\varphi$ is the Coriolis parameter; Ω is the angular speed of the earth's rotation; φ is latitude; $m = \frac{1}{s}\frac{1+\sin 60°}{1+\sin\varphi}$ is a scale factor; s is the factor by which the scale is increased at the 60°N parallel; R is the gas constant; $\lambda = \frac{c_p - c_v}{c_p}$; c_p and c_v are specific heats at constant pressure and volume, respectively; $\Phi = gz$ is geopotential; z is altitude; g is the acceleration of gravity; θ is potential temperature; p is pressure; $P = 1000$ mb; ε is the heat influx per unit volume.

The boundary conditions in the vertical direction are taken in the form

$$\omega = 0 \qquad \text{for} \qquad p = 0,$$

$$w = \frac{dz}{dt} = 0 \qquad \text{for} \qquad p = P \tag{6}$$

or

$$\frac{d\Phi}{dt} = \frac{\partial \Phi}{\partial t} + m\left(u\frac{\partial \Phi}{\partial x} + v\frac{\partial \Phi}{\partial y} + \omega\frac{\partial \Phi}{\partial p}\right) = 0. \tag{7}$$

Condition (7) can be expressed, as equations (1), (2) and (4) and with the aid of equation (3), in the divergent form

$$\frac{\partial \Phi}{\partial t} + m\left(\frac{\partial u\Phi}{\partial x} + \frac{\partial v\Phi}{\partial y} + \frac{\partial \omega\Phi}{\partial p}\right) = 0. \tag{8}$$

The following conditions are assumed at lateral boundaries Γ: the normal component V_n of the velocity vector and the derivative along the normal n to the lateral boundary of the tangential velocity vector component $V(\tau)$ vanish, i. e.,

$$V_n\big|_\Gamma = 0, \tag{9}$$

$$\frac{\partial V\tau}{\partial n}\bigg|_\Gamma = 0. \tag{10}$$

If boundary Γ passes along the equator, then $\Gamma = 0$, and as a result of condition (9) we have

$$\frac{\partial \Phi}{\partial n}\bigg|_\Gamma = 0. \tag{11}$$

The initial conditions are given by

$$u = u(0), \quad v = v(0), \quad \Phi = \Phi(0).$$

Here, due to the unavailability of the required information on the real wind, we take as initial velocities their geostrophic values:

$$u_g = -\frac{m}{l}\frac{\partial \Phi}{\partial y}, \quad v_g = \frac{m}{l}\frac{\partial \Phi}{\partial x}.$$

Replacing θ in equation (4) by means of equation (5) gives

$$\frac{\partial}{\partial t}\frac{\partial \Phi}{\partial p} + m\left(\frac{\partial}{\partial x}u\frac{\partial \Phi}{\partial p} + \frac{\partial}{\partial y}v\frac{\partial \Phi}{\partial p} + \frac{\partial}{\partial p}\omega\frac{\partial \Phi}{\partial p} + \alpha\omega\frac{\partial \Phi}{\partial p}\right) + \frac{R}{pc_p\rho}\varepsilon = 0$$

or

$$\frac{\partial}{\partial t}\frac{\partial \Phi}{\partial p} + m\left[\frac{\partial}{\partial p}\left(\frac{\partial u\Phi}{\partial x} + \frac{\partial v\Phi}{\partial y} + \frac{\partial \omega\Phi}{\partial p}\right) + \alpha\omega\frac{\partial \Phi}{\partial p} - \right.$$

$$\left. - \left(\frac{\partial \Phi}{\partial x}\frac{\partial u}{\partial p} + \frac{\partial \Phi}{\partial y}\frac{\partial v}{\partial p} + \frac{\partial \Phi}{\partial p}\frac{\partial \omega}{\partial p}\right)\right] + \frac{R}{pc_p\rho}\varepsilon = 0, \tag{12}$$

where

$$\alpha = \frac{1-\lambda}{p}.$$

These equations were approximated by finite-difference expressions employing a staggered space and time grid, used previously in /4/.

These equations were integrated over the gridded domain inside a square, the vertices of which are located at points on the equator with longitudes 0, 90, 180°E and 90°W on a polar stereographic projection map, i. e., the forecast pertains to the major part of the Northern Hemisphere. The horizontal incremental step was 300 km and the time step 15 mins. Calculations were made for the 100, 300, 500, 700, 900 and 1000 mb surfaces.

TABLE 1

Starting date	Surface (mb)	One-day forecast			Three-day forecast		
		σ	$E \cdot 100$	$r \cdot 100$	σ	$E \cdot 100$	$r \cdot 100$
28 May 1967	100	4.9	96	51	8.3	87	63
	300	10.0	85	55	11.9	67	76
	500	4.7	86	64	8.3	66	76
	700	3.7	82	67	6.9	71	73
	900	3.5	83	61	5.9	73	71
	1000	4.2	78	64	6.3	69	74
28 June 1967	100	4.3	91	63	6.6	92	64
	300	6.5	75	67	9.7	74	68
	500	4.8	74	68	7.1	76	65
	700	3.7	73	72	6.1	79	64
	900	3.3	77	65	5.6	80	60
	1000	4.5	83	57	7.1	81	60
25 March 1968	100	6.4	92	70	13.0	90	67
	300	5.8	52	86	11.3	70	75
	500	6.2	75	73	9.4	72	77
	700	4.8	76	73	8.0	82	74
	900	3.9	65	78	7.8	89	71
	1000	5.4	67	77	8.4	78	69

The adiabatic model was used to compute forecasts for up to three days. Table 1 lists the standard error σ, relative error E and correlation coefficient r for three forecasts for one and three days. It follows from the tabulated data that the best forecasts are for the 300 and 500 mb surfaces, while the poorest are for the 100 mb surface.

The above data show that this forecasting model provides satisfactory results and can be used in experiments which make allowance for nonadiabatic factors.

CALCULATION OF HEAT INFLUX

In the nonadiabatic forecasting model the heat influx should be calculated at each time step, i. e., it should be forecast together with other meteorological variables, which involves great difficulties. However, to analyze the effect of heat influx as weather-forming factors they can be precalculated, irrespective of the forecasting model, and then they can be introduced into this model at the appropriate time steps. Such a technique was employed previously, for example, in /2, 20, 21, 23/ in forecasting models for different time periods.

This technique was used also in this study when analyzing the effect of heat influxes on variation in the geopotential for three days. In experiments with numerical weather forecasting the heat influx was calculated as an average for each of the days from the daily average values of meteorological variables. Here the effective mean value of the sine of the solar altitude h, needed for calculating short-wave radiation, was calculated at each point from the expression

$$\overline{\sin h} = \frac{\int_0^{t_0} \sin\, dt}{\int_0^{t_0} dt},$$

where t_0 is the duration of daylight.

The radiative heat fluxes and influxes ε_r were calculated by the method of /1–3, 5, 8/. The radiative fluxes were calculated for eleven levels (1000, 900, ..., 0 mb). The heat influxes, on the other hand, were determined from the difference in the fluxes for layers corresponding to the forecasting model (1000–900, 900–700, 700–500, 500–300 and 300–100 mb). The cloudiness was specified as a single layer, the upper and lower boundaries of which were referred to some of the given levels. In the case of N-point cloudiness the heat influx was calculated from the expression

$$\varepsilon_r^N = \varepsilon_r^{10} \frac{N}{10} + \varepsilon_r^0 \left(1 - \frac{N}{10}\right),$$

where ε_r^{10} and ε_r^0 are the radiative heat influxes calculated with 10-point cloudiness and with clear skies, respectively.

The heat influxes were calculated from actual data on temperature and cloudiness; information on the cloud cover (number of points) and its altitude was supplemented by satellite data.

Data on the air humidity outside the cloud layers were re-presented by their climatic data, while for the cloud layers it was assumed that there is complete saturation at the corresponding air temperature.

Since the radiative heat influx acts simultaneously with its eddy counterpart, an attempt was made to take into account simultaneously these two forms of heat influx. This was done in several ways. In one, use was made of an idea suggested earlier in /6, 7, 12/ concerning singling out the standard portions of the heat influx. We assume that $\varepsilon_r = \bar{\varepsilon}_r + \varepsilon_r'$ and that $\varepsilon_e = \bar{\varepsilon}_e + \varepsilon_e'$, where ε_r and ε_e are the radiative and eddy heat influxes, $\bar{\varepsilon}_r$ and $\bar{\varepsilon}_e$ are their standard values, while ε_r' and ε_e' are deviations from it. The standard values satisfy the condition $\bar{\varepsilon}_r + \bar{\varepsilon}_e = 0$, which makes it possible to calculate the standard temperature; this can be obtained sufficiently close to the climatic value as, for example, in /5/, with the calculations carried out for the cloudless atmosphere. On the other hand, allowance for cloudiness would have complicated determination of the standard temperature, which also means the standard heat influx. At the same time it is known that cloudiness is a governing factor in the formation of radiative heat influx /19/. Hence we took as standard value of the radiative heat influx the value calculated from climatic values of the temperature, humidity, cloudiness and albedo of the surface. The variations in atmospheric pressure are thus determined by values of ε_r' and ε_e', calculated from real values of the meteorological variables.

An attempt was also made to take into account directly the eddy influx of heat in the boundary layer (1000–900 mb). In this case it was assumed that there is no eddy heat influx H at the upper boundary of the boundary layer, and the eddy heat influx at the surface was determined by the balance equation, which in the absence of flux in the soil and of evaporation has the form $H + Q = 0$, where Q is the radiation balance of the surface. The idea of mutual compensation of standard heat influxes can be implemented also in this case.

We have thus carried out calculations to make allowances for nonadiabatic factors in four ways: I — only the radiative heat influx was considered (ε_r); II — only that part of the radiative influx was considered which is not compensated by the standard part of the eddy heat influx $(\varepsilon_r' = \varepsilon_r - \bar{\varepsilon}_r)$; III — allowance was made for the radiative heat influx and for the eddy influx of heat in the boundary layer; IV — consideration was given to noncompensated parts of the radiative heat influx and of the eddy heat influx in the boundary layer.

EXPERIMENTAL RESULTS

The difference in forecast values of the geopotential obtained by the nonadiabatic and adiabatic models is the contribution of nonadiabatic factors to the variation in the geopotential. The mean algebraic value of the contributions of nonadiabatic factors to the geopotential for three days for the major part of the Northern Hemisphere for the four versions are presented in Table 2.

Consider now results obtained by completely taking into account the radative heat influx (version I). It follows from the table that the radiative heat influx results on the average in a small increase in geopotential (up to 0.94 decameters for three days) of the lower isobaric surfaces and to a substantial reduction in geopotential (up to 11.47 decameters for three days) of the upper isobaric surfaces. Actually no such large systematic reduction in the geopotential of upper surfaces is observed, since the radiative heat influx should be compensated by the eddy flux. This is confirmed well by data obtained using version II, when allowance is made only for the noncompensated part of the radiative heat influx. The maximum (of the average) value of contributions of heat influx in this version amounts to approximately 2.07 decameters for three days. Here, near the earth's surface the different contributions have different signs in different situations.

The average contributions of nonadiabatic factors according to version III (allowance for radiative heat influx and approximate incorporation of eddy heat influx in the boundary layer) were found to be smaller than for version I and larger than for version II. It is characteristic that, as compared with version I, the negative contributions to the variation in the geopotential of upper surfaces decreased. This points to the fact that the suggested approximate method of allowing for turbulence in the boundary layer generally reflects properly the process. However, the relatively high values of these contributions show that in this version eddy heat transfer is not as yet completely incorporated. The contributions of nonadiabatic factors were found to be smallest (less than 1.25 decameters) according to version IV and to have different signs at different levels.

The contributions of nonadiabatic factors and the variation in geopotential depend markedly on latitude. Thus, for example, the contributions of nonadiabatic forces to variations in the geopotential of the 100 mb surface according to version II for summers in tropics amount to 2—6 decameters, while in polar regions the range is from -2 to +3 decameters. It should, however, be emphasized that the isolines of different contributions are not

TABLE 2

Starting date	Surface (mb)	Average contribution of heat influx to the variation in geopotential for version				Average values of forecasting errors for version					
		I	II	III	IV	0	I	II	III	IV	V
25 March 1968	100	—11.47	1.98	—10.88	0.60	3.40	—8.07	5.38	—7.48	4.00	3.99
	300	—7.34	1.23	—6.78	—0.16	—0.28	—7.62	0.95	—7.06	—0.44	0.28
	500	—4.45	1.39	—3.91	0.00	—3.16	—7.61	—1.77	—7.07	—3.16	—2.62
	700	—0.40	1.17	0.10	—0.22	—3.12	—3.52	—1.95	—3.02	—3.34	—2.62
	900	0.19	0.20	0.56	—1.19	—3.50	—3.31	—3.30	—2.94	—4.69	—3.13
	1000	0.38	0.20	1.40	—0.25	—2.42	—2.04	—2.22	—1.02	—2.18	—1.40
28 May 1967	100	—7.55	0.88	—3.57	—0.05	—2.47	—10.02	—1.59	—6.04	—2.52	1.51
	300	—4.23	0.33	—0.33	—0.60	—1.97	—6.20	—1.64	—2.30	—2.57	1.93
	500	—2.07	0.45	1.91	—0.48	—0.86	—2.93	—0.41	1.05	—1.34	3.16
	700	0.88	1.46	4.83	0.55	—1.87	—0.99	—0.41	2.96	—1.34	2.08
	900	—0.04	—0.22	3.93	—1.13	—1.15	—1.19	—1.37	2.78	—2.30	2.82
	1000	0.03	—0.06	0.42	—0.01	—1.64	—1.61	—1.70	—1.22	—1.65	—1.25
28 June 1967	100	—6.21	1.88	—1.70	1.01	—1.81	—8.02	**0.07**	—3.51	—0.80	2.70
	300	—4.03	1.16	0.48	0.30	—0.66	—4.69	0.50	—0.18	—0.36	3.80
	500	—1.73	0.95	2.43	0.08	—0.91	—2.64	0.04	—1.52	—0.83	3.25
	700	0.94	2.07	5.40	1.19	—0.76	0.18	1.31	4.64	0.43	3.70
	900	—0.19	—0.37	4.26	—1.25	—0.52	—0.71	—0.89	3.74	—1.77	3.93
	1000	0.08	0.07	0.33	0.11	—1.38	—1.30	—1.31	—1.05	—1.27	—1.13

parallel to latitude circles, but have a very complicated shape. Such a sharp variation in contributions is observed depending on the season. Thus, for example, the contributions of the radiative heat influx to the variation in the geopotential of the 500 mb surface in July in tropics and polar regions is from 0 to -2 decameters, in March in the tropics it ranges from -2 to -4 decameters, and in polar regions from -5 to -7 decameters. We also note that in individual cases the difference in contributions of nonadiabatic factors at adjoining points for the same synoptic situation is as high as 2–3 decameters. These data point to the very complicated nature of the dependence of heat-influx contributions to the variation in geopotential on the physicogeographic conditions, as well as on the specific synoptic situation.

We now consider the effect of nonadiabatic factors on the forecast of the general background pressure. Table 2 lists the mean algebraic errors in forecasts for three days, calculated over the major part of the Northern Hemisphere for all four versions. In addition, the average errors obtained from the adiabatic model (version 0) are tabulated, as well as for version V in which only eddy heat influx in the boundary layer is taken into account.

TABLE 3

Surface (mb)	25 March 1968				28 May 1967				28 June 1967			
	I	II	III	IV	I	II	III	IV	I	II	III	IV
ΔE												
100	−10	−6	−7	−2	−28	0	−25	−3	−45	−1	−12	−2
300	−14	−3	−12	−2	−6	−1	−1	−2	−8	−1	−1	−1
500	−16	0	−13	−1	−3	0	0	−1	−4	4	2	4
700	−3	4	−2	0	0	0	−4	0	0	−5	−20	−2
900	0	2	0	−5	−2	−2	−8	−5	0	−1	−14	−2
1000	1	0	2	1	0	0	−2	−2	1	1	1	0
Δr												
100	1	−1	3	0	−5	0	−7	0	13	−1	−3	−1
300	−2	−2	−1	−2	−1	0	0	−1	−1	−1	−1	−1
500	−1	−1	1	−1	0	0	0	0	0	3	3	4
700	−1	1	0	1	−1	−1	0	−1	0	−3	−2	−3
900	−1	0	−1	0	−1	0	−1	0	−1	0	0	0
1000	0	−1	−1	0	0	0	−3	−2	0	0	1	−1

It should be noted that for the version in which eddy heat influx in the boundary layer is taken into account in a simplified manner (III–V) the forecast at sea level almost always has a large error. At the same time the greatest reduction in mean errors at all levels (in 12 out of 18) occurs for version II. Consequently, from the standpoint of forecasting the general pressure background, this version is the best and yields only slightly better results compared with the adiabatic forecast.

However, from the viewpoint of forecasting quality based on the relative error E and correlation coefficient r the quality of non-adiabatic forecasts for all the above versions remains on the level of adiabatic forecasts. This is seen from data of Table 3, which lists values of ΔE and Δr, characterizing the difference in the relative errors and correlation coefficients for nonadiabatic and adiabatic forecasts (positive values of ΔE and Δr correspond to an improvement in nonadiabatic as compared with adiabatic forecasts).

It is likely that a more complete allowance for eddy heat transfer with radiative transport, as well as differentiated consideration of the mutual compensation of these two kinds of heat transfer in different regions and at different vertical levels, can improve the estimates of nonadiabatic processes.

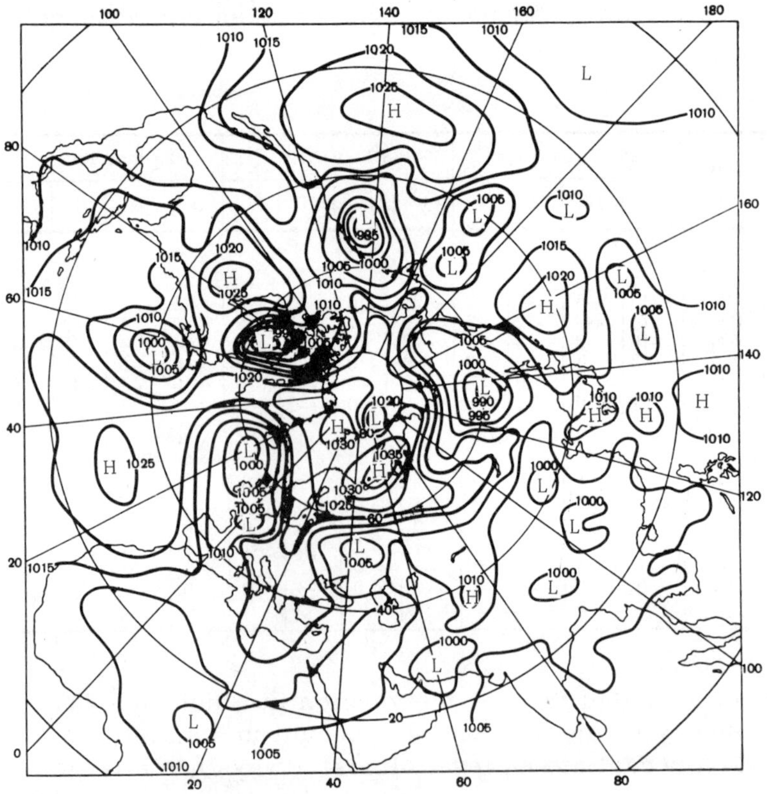

FIGURE 1. Surface pressure map for 0300 hrs on 28 May 1967.

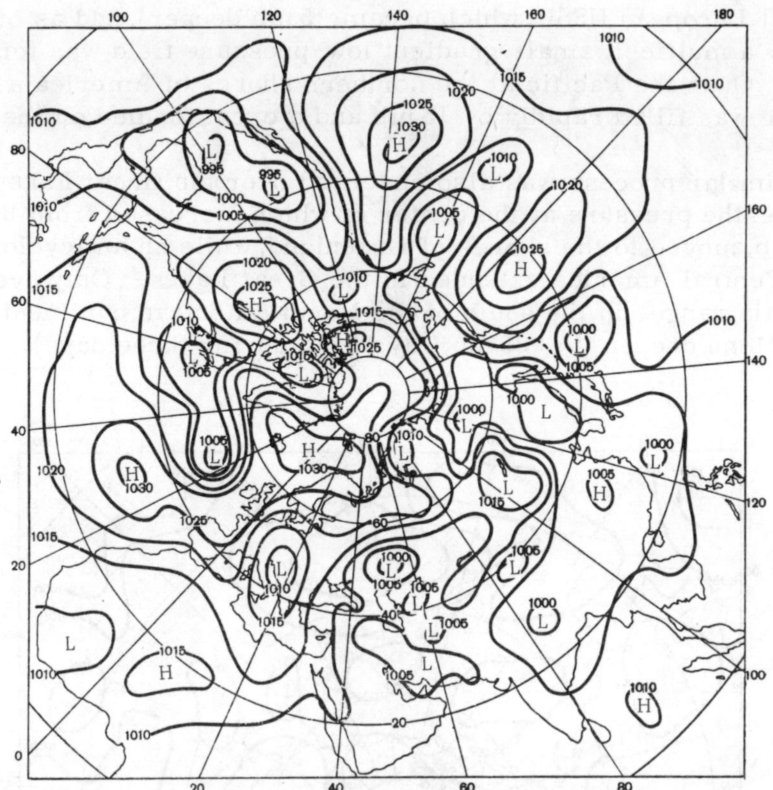

FIGURE 2. Surface pressure map for 0300 hrs on 31 May 1967.

In conclusion we consider the results of computing a nonadiabatic forecast (version II) for three days using the data for 28 May 1967. The characteristic process over the Northern Hemisphere from 28 to 31 May 1967 was the filling of cyclones over the entire area, without significant anticyclonic development. Both at the surface and in the middle troposphere the pressure gradients became smaller and the air flows in moderate latitudes were oriented zonally. In the middle latitude of the Atlantic and Pacific, as well as over Southern Europe the baric gradients at the 500 mb level in the high-frontal zone decreased for the three days from 32 to 20 decameters per 1000 km.

Over the North Atlantic a heavy high cyclone was being filled, while an anticyclone over Western Europe was slightly reinforced. An anticyclonic region extended from this anticyclone to the surface from the southwest at the South Atlantic (Figures 1 and 2).

Over Western Europe and Asia cyclones were filled primarily at the surface by 5−10 mb (with the exception of the cyclone over

Central European USSR, which became 5 mb deeper), and as of
31 May a multicell small-gradient low-pressure field was formed
there. Over the Pacific at the northern shores of America a
cyclone was filled rapidly by 15 mb and a crest formed on the third
day.

A similar process was also observed over the Great Lakes. The
cyclone, the pressure at the center of which increased from 990 to
1015 mb, moved to the shores of Greenland, while an anticyclone
from Central America extended to the Great Lakes. Only over
mountain ranges in the south of the North American continent did
the cyclone deepen (to 995 mb; by 15 mb during three days).

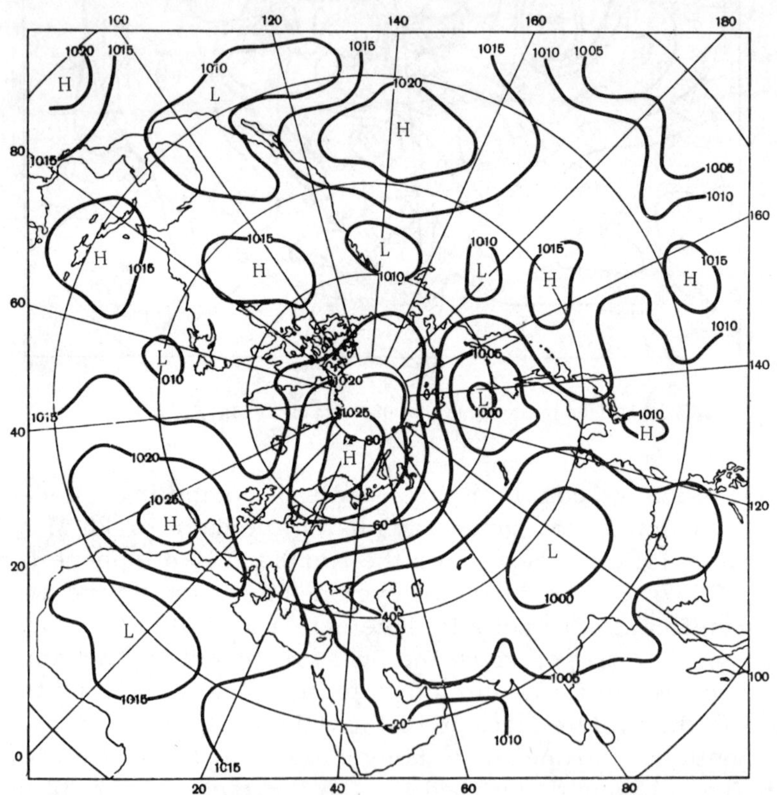

FIGURE 3. Forecasting pressure map for 0300 hrs on 31 May 1967.

The forecasting map shown in Figure 3 reflects (to an even
greater extent than is actually true) the weakening of the activity
of atmospheric processes. The principal pressure centers of the

actual and forecasting maps coincide, however, their intensity on the latter is approximately 5–10 mb lower than the actual value. The high-altitude forecasting maps are closer to the actual maps.

BIBLIOGRAPHY

1. B e l o v , P.N. Karty radiatsionnogo balansa i vozmozhnosti ikh ispol'zovaniya v sinopticheskoi praktike (Radiation Balance Maps and the Feasibility of their Utilization in Synoptic Practice).— Meteorologiya i Gidrologiya, No.4. 1967.

2. B e l o v , P.N. and R.L.A l p a t o v a. Nekotorye rezul'taty chislennykh eksperimentov po uchetu radiatsionnogo pritoka tepla v chislennom prognoze (Some Results of Numerical Experiments in Making Allowance for Radiative Heat Influx in Numerical Forecasting).— Trudy Gidrometeorologicheskogo Tsentra SSSR, No.11. 1967.

3. B e l o v , P.N. and A.F.K i v g a n o v. Ob izmeneniyakh temperatury i geopotentsiala, obuslovlennykh radiatsionnym pritokom tepla (On Temperature and Geopotential Variations Induced by Radiative Heat Influx).— Trudy MMTs, No.8. 1965.

4. B e r k o v i c h , L.V. Trekhurovennaya skhema prognoza meteoelementov po polnym uravneniyam (A Three-Level Model for Forecasting Meteorological Parameters on the Basis of General Equations).— Trudy MMTs, No.14. 1966.

5. B e r k o v i c h , L.V. O sovmestnom uchete radiatsionnogo i turbulentnogo pritokov tepla v chislennom prognoze (On Combined Consideration of Radiative and Eddy Heat Influxes in Numerical Forecasting).— Trudy GMTs, No.11. 1967.

6. B l i n o v a , E.N. K gidrodinamicheskoi teorii klimata dolgosrochnogo prognoza pogody (The Hydrodynamic Theory of Climate and of Long-Range Weather Forecasting).— In Sbornik: "Gidrodinamicheskii dolgosrochnyi prognoz pogody." Mozkva, Izdatel'stvo Nauka. 1964.

7. B l i n o v a , E.N. Obobshchenie gidrodinamicheskoi teorii klimata i dolgosrochnogo prognoza pogody (Generalization of the Hydrodynamic Theory of Climate and of Long-Range Weather Forecasting).— Trudy Gidrometeorologicheskogo Tsentra SSSR, No.15. 1967.

8. V o i t k o , N.V. and A.F.K i v g a n o v. Raschet i analiz radiatsionnykh potokov i pritokov tepla pri nalichii oblachnosti (Calculation and Analysis of Radiative Heat Fluxes and Influxes Under Cloudy Skies).— Trudy MMTs, No.11. 1966.

9. G u t s h a b a s h , S D. Vychislenie radiatsionnykh potokov v oblachnoi i bezoblachnoi atmosfere (Calculation of Radiative Fluxes Under Cloudy and Clear Skies).— Meteorologiya i Gidrologiya, No.12. 1969.

10. D m i t r i e v a - A r r a g o , L . R. Raschet potokov i pritokov dlinnovolnovoi radiatsii v oblachnykh usloviyakh (Calculation of Long-Wave Radiative Fluxes and Influxes Under Cloudy Conditions).— Trudy GGO, No.197. 1968.

11. D m i t r i e v a - A r r a g o , L.R. et al. O skheme rascheta potokov korotkovolnovoi radiatsii (On a Model for Calculating Fluxes of Short-Wave Radiation).— Meteorologiya i Gidrologiya, No.8. 1970.

12. K i b e l ' , I.A. Vvedenie v gidrodinamicheskie metody kratkosrochnogo prognoza pogody (Introduction to Hydrodynamic Methods of Long-Range Weather Forecasting). Moskva, Gostekhizdat. 1957.

13. K o n d r a t ' e v , K.Ya., I.L.V u l i s, and G.A.N i k o l ' s k i i. Ob uchete radiatsionnykh faktorov pri modelirovanii protsessov obshchei tsirkulyatsii atmosfery (On Making Allowance for Radiation Factors in Simulating Processes of the General Circulation of the Atmosphere).— Meteorologiya i Gidrologiya, No.7. 1969.

14. M a r c h u k, G.I. Teoreticheskaya model' prognoza pogody maloi zablagovremennosti (A Theoretical Model of Short-Range Weather Forecasting).— Izvestiya AN SSSR, Seriya Geofizicheskaya, No.5. 1964.

15. M a s h k o v i c h, S.A. K uchetu neadiabaticheskikh vozdeistvii pri chislennom kratkosrochnom prognoze davleniya (Incorporation of Nonadiabatic Effects in Numerical Short-Range Pressure Forecasting).— Trudy TsIP, No.126. 1963.

16. M e l e s h k o, V.P. K voprosu o chislennom prognoze oblachnosti i osadkov s uchetom transformatsii vozdushnykh mass (Concerning Numerical Forecasting of Cloudiness and Precipitation with Allowance for Air-Mass Transformation).— Trudy GGO, No.151. 1964.

17. P u r g a n s k i i, V.S. K uchetu vliyaniya neadiabaticheskikh faktorov pri kratkosrochnom prognoze polei geopotentsiala na nizhnikh urovnyakh atmosfery (Incorporation of the Effect of Nonadiabatic Factors in Short-Range Forecasting of the Geopotential Field in Lower Atmospheric Levels).— Trudy TsIP, No.126. 1963.

18. R a k i p o v a, L.R. and B.E.S h n e e r o v. Uchet radiatsionnykh pritokov tepla v zadachakh dinamicheskoi meteorologii (Consideration of Radiative Heat Influx in Problems of Dynamic Meteorology).— In Sbornik: "Meteorologicheskie issledovaniya," No.16. Moskva, Izdatel'stvo "Nauka." 1968.

19. F e i g e l ' s o n, E.M. Luchistyi teploobmen i oblaka (Radiative Exchange and Clouds). Leningrad, Gidrometeoizdat. 1970.

20. S h v e t s, M.E. and B.E.S h n e e r o v. K uchetu rezul'tatov radiatsionnykh izmerenii so sputnikov v neadiabaticheskoi modeli atmosfernykh dvizhenii (Incorporation of Satellite Radiation Measurements in the Nonadiabatic Model of Atmospheric Motions).— DAN SSSR, Vol.152, No.3. 1963.

21. L e i t h, C.E. Numerical Simulation of the Earth's Atmosphere.—Methods in Computational Physics, Vol.4. 1965.

22. M a n a b e, S. and R.F.S t r i c k l e r. Thermal Equilibrium of the Atmosphere with a Convective Adjustment.— J. Atmos. Sci., Vol.21, No.4. 1964.

23. M i n t z, Y. Very Long-Term Global Integration of the Primitive Equations of Atmospheric Motion.— WMO Technical Note No.66. Geneva. 1965.

24. M i y a k o d a, K., J.S m a g o r i n s k y, R.F.S t r i c k l e r, and G.D.H e m b r e e. Experimental Extended Predictions with a Nine-Level Hemispheric Model.— Monthly Weather Review, Vol.97, No.1. 1969.

25. S m a g o r i n s k y, J., S.M a n a b e, and J.H o l l o w a y. Numerical Results from a Nine-Level General Circulation Model of the Atmosphere.— Monthly Weather Review, Vol. 93, 1965.

26. T a k a h a s h i, K., A.K a t a y a m a, and T.A s a k u r a. A Numerical Experiment of the Atmospheric Radiation.— J. Met. Soc. Japan, Ser.II, Vol.38, No.4. 1960.

UDC 551.509

SOME ASPECTS OF THE SOLUTION OF INVERSE PROBLEMS OF SATELLITE METEOROLOGY

E. E. Tolstikov

The paper presents classes of synoptic situations obtained in the course of considering the feasibility of relating radiation temperature fields to synoptic situations. Alphabets of these classes are given and examples are examined of determining classes of synoptic situations from the statistical structure of radiation temperature fields.

Analysis of actinometric information supplied by meteorological satellites has enabled the following problem to be formulated: how to create a classification of synoptic processes which would allow one to recognize a given synoptic situation from the statistical structure of the radiation-temperature fields over a given locality /1/. In the course of studying statistical characteristics of radiation-temperature fields, the feasibility was also examined of selecting numerical values of the calculated characteristics of these fields as attributes of classes, and their probability characteristics were determined. A determination was thus made of an alphabet of classes (description of classes and occurrences), which serves as a point of departure for utilizing some criterion of recognition theory for determining classes of synoptic situations. The boundaries of classes were determined by a scheme for calculating a priori probabilities, most completely reflecting the essence of recognition (the ideal observer criterion). This method was employed to construct classes for which statistical characteristics of the radiation temperature fields were calculated and studied. These calculations were carried out on the assumption of homogeneity of radiation temperature fields for each type of cloudiness. Classes of synoptic situations, their attributes and gradations were obtained for cloudiness types determined by nephanalysis.

The data used in our study of the statistical characteristics of radiation fields were those obtained by the Cosmos 144 satellite.

The purpose of this study was the investigation of possible classes of synoptic situations, their attributes and gradations. Alphabets were composed for the individual classes and examples were examined in which synoptic situations were singled out from statistical characteristics of the field of departing radiation. Here estimates were made of computational errors, arising as a result of both the finite nature of the recording interval and the replacement of integration by summation.

These studies led to four classes of synoptic situations, which fully reflect the synoptic state of the region under study (160°E–120°W, 50 – 70°N) for May and June.

Class I – A_I. To the south and north (less frequently to the east and west) of the region under study high-pressure regions are located with a prevalence of middle clouds or clear skies. Here a frontal cloudiness band or the center of a filling low are almost invariably observed in the region under study.

Class II – A_II. A col, frequently accompanied by diffuse fronts, predominates in this region. Sometimes a cyclone with a large warm sector is observed instead of the col. The cloudiness within the col and the warm sector of the cyclone is discontinuous, consisting of cloud bands, and the cloudiness is not dense. There is a prevalence of altocumulus (middle clouds) and cirrus (high clouds), sometimes covering the entire sky.

Class III – A_III. Multicell depressions or extensive cyclonic regions prevail over the region under study. Cumulus (cellular) cloudiness prevails but stratus, which is associated with the invasion of stable Arctic masses, is also observed sometimes.

Class IV – A_IV. The northern part of the Pacific anticyclone, having at its periphery arctic-front cyclones, prevails over the region. Substantial high and middle clouds and individual bands of low cloudiness, mainly of cumulus (cellular) form, predominate here.

The following were selected as indicators for compiling the alphabet of classes: dimensionless recording length $X_1=\alpha T$, which reflects both the micro- and macrostructure of fields of departing radiation (T is the recording length or the interval of homogeneity of the radiation-temperature field determined by the cloud-cover structure; α is the correlation factor, whose dimension is the reciprocal of length), and the mathematical expectation of the radiation temperature field $X_2=\overline{t}$. Each indicator has three gradations ($R = 3$): for dimensionless recording length $x'_1 < 7$, $x'_2 > 7-9$, $x'_3 > 9$; for radiation temperatures $x^2_1 > (-10°)$, $x^2_2 = (-10°)$ to $(-20°)$, $x^2_3 < (-20°)$.

It is extremely difficult to construct an alphabet immediately for four synoptic situations, and hence, applying the method of successive classification, we composed at the first stage an alphabet for classes A' and A'' such that $A' \ni A_\mathrm{I} \bigcup A_\mathrm{II}$ and $A'' \ni A_\mathrm{III} \bigcup A_\mathrm{IV}$. The probabilities of appearance of these classes under actual conditions are $P(A') = 0.575$ and $P(A'') = 0.425$.

The alphabet consists of an ensemble of occurrences b_j, a probability distribution of reference occurrences b_j pertaining to classes A' and A'', i. e. $P(b_j/A'')$ and $P(b_j/A'')$ respectively, the probability $P(b_j)$ of occurrence of b_j under actual conditions, as well as the a posteriori probability distributions $P(A'/b_j)$ and $P(A''/b_j)$ of hypotheses concerning the belonging of occurrence b_j to classes A' and A''.

At the second stage we constructed alphabets for classes A_I and A_II and classes A_III and A_IV. As a result, alphabets of four classes of synoptic situations were established for the purpose of referring the occurrence to one of these classes. Table 1 lists the alphabets for classes A' and A'' as well as A_I and A_II.

The alphabet of classes used in line with the ideal observer criterion makes it possible to calculate examples of the determination of a class of synoptic situations from the statistical characteristics of the radiation temperature field. For example, such a field obtained on 21 May 1967 comprises occurrence b_5 $(\alpha T = 7.8,\ t = -15°)$. This occurrence belongs to class A' with probability $P(A'/b_5) = 0.700$ and to class A_II with probability $P(A_\mathrm{II}/b_5) = 0.806$. Consequently, the radiation temperature field for 21 May 1967 corresponds to a class which can be described as follows: a col prevails in the region under study, often together with diffuse fronts, or a cyclone is observed with a large warm sector. The actual map (Figure 1) shows in the western part of the region under study the center of a cyclone with a 1,000 mb isobar. A large warm section of this sector is located in the southern part of this region. Low-cloudiness weather with individual cellular cloud bands predominates here. The class of synoptic situation was hence determined correctly.

To increase the size of the sample of occurrences and to check the correctness of the above method, we considered additional data; the results of their processing are described below.

The selection of data is always a complicated and difficult problem. This is best illustrated by the selection of necessary information from the Cosmos 144 satellite. However the limitations (a specific region of the globe; to eliminate the daily variation in radiation the information was selected within the range of ± 4 hours about noon, while the yearly variation in radiation was

TABLE 1

b_j	b_1	b_2	b_3	b_4	b_5	b_6	b_7	b_8	b_9
				Classes A' and A''					
$X_1 = \alpha T$	<7	<7	<7	$7{-}9$	$7{-}9$	$7{-}9$	>9	>9	>9
$X_2 = \bar{t}$	$>(-10)$	$(-10){-}(-20)$	$<(-20)$	$>(-10)$	$(-10){-}(-20)$	$<(-20)$	$>(-10)$	$(-10){-}(-20)$	$<(-20)$
$P(b_j/A')$	0.000	0.267	0.134	0.066	0.335	0.066	0.000	0.066	0.066
$P(b_j/A'')$	0.000	0.000	0.000	0.092	0.182	0.092	0.272	0.362	0.000
$P(b_j)$	0.000	0.154	0.078	0.078	0.270	0.078	0.112	0.192	0.038
$P(A'/b_j)$	—	1.000	1.000	0.480	0.700	0.480	0.000	0.200	1.000
$P(A''/b_j)$	—	0.000	0.000	0.520	0.300	0.520	1.000	0.800	0.000
				Classes A_{I} and A_{II}					
$X_1 = \alpha T$	<7	<7	<7	$7{-}9$	$7{-}9$	$7{-}9$	>9	>9	>9
$X_2 = \bar{t}$	$>(-10)$	$(-10){-}(-20)$	$<(-20)$	$>(-10)$	$(-10){-}(-20)$	$<(-20)$	$>(-10)$	$(-10){-}(-20)$	$<(-20)$
$P(b_j/A_{\mathrm{I}})$	0.000	0.500	0.133	0.000	0.167	0.000	0.000	0.000	0.000
$P(b_j/A_{\mathrm{II}})$	0.000	0.111	0.000	0.111	0.445	0.111	0.000	0.111	0.111
$P(b_j)$	0.000	0.267	0.133	0.067	0.334	0.067	0.000	0.066	0.066
$P(A_{\mathrm{I}}/b_j)$	—	0.750	1.000	0.000	0.194	0.000	—	0.000	0.000
$P(A_{\mathrm{II}}/b_j)$	—	0.250	0.000	1.000	0.806	1.000	—	1.000	1.000

eliminated by using information for May and June), which are needed in order to select homogeneous data over the area under study, and the need for integrated utilization of TV, infrared and atinometric information and ordinary meteorological information, made the selection of information for subsequent studies particularly difficult.

We studied information obtained from Cosmos 156, 184, 206 and 226, and Meteor 1 and 5 satellites. Information from Cosmos 156 and 206 had to be rejected. Cosmos 184 did not provide information for June. The Cosmos 226 satellite passed the region under study at night, which made it impossible to select homogeneous information. Meteor 5 started functioning in June 1970, while information from it is available only as of 28 June 1970. Only the Meteor 1 satellite provided information for June 1969, which to some extent corresponds to conditions of data selection.

Fifty passes of the Meteor 1 satellite were over the area under study in June 1969, but due to restrictions imposed on the selection of information only seven of these fifty could be used.

An additional difficulty was the determination of the correction for readings of the Meteor 1 satellite. The alphabet of classes was constructed for information supplied by Cosmos 144. It was necessary to determine the differences between the information supplied by these two satellite series. For this we compared readings for the same day and hour over the open sea, ice, snow, snow-covered mountains, etc. It was found that the departing radiation measured from Cosmos 144 and Meteor 1 differ by approximately 10°. Thus, if the radiation-temperature field obtained from Meteor 1 is corrected by this amount, then the statistical characteristics of radiation fields can be described by our alphabet.

We have studied radiation-temperature fields obtained over the region under study for 1, 4, 7–11 June 1969. Using the above correction we calculated statistical characteristics which are indicators of the classes. Below are presented the indicators (lines 1 and 2), corresponding occurrences (line 3), and classes of synoptic situations to which these occurrences are referred.

Number	1	4	7	8	9	10	11
aT	7.54	7.27	9.82	7.00	6.55	8.04	6.16
$\bar{t}$	−9.50	−14.10	−22.12	−7.80	−20.60	−17.67	−15.05
b_j·	b_4	b_5	b_9	b_4	b_3	b_5	b_2
A_i	A_{III}	A_{II}	A_{II}	A_{III}	A_I	A_{II}	A_I

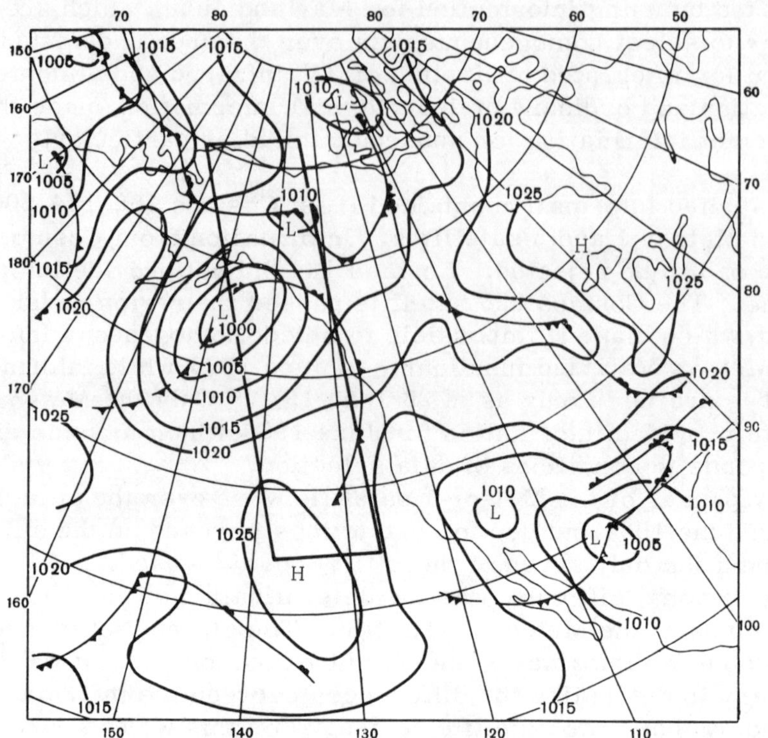

FIGURE 1. Synoptic map for 0300 hrs on 21 May 1967. The region under study is outlined on the map.

According to the table, the radiation temperature field for 1 June corresponds to class A_{III}. Actually a series of cyclones passed over this region at a moderate front, and a rear anticyclone, formed in a stable Arctic mass with a 1020 mb isobar at the center, started to develop. Consequently, the radiation temperature field for 1 June actually corresponds to class A_{III}.

The radiation temperature field for 4 June corresponds to class A_{II}. We see on the synoptic map that at the time of the satellite observation a col prevailed over the selected region, i. e., the radiation field actually expresses class A_{II}.

The radiation field for 7 June was classified (using the ideal observer criterion) also as class A_{II}. The synoptic map indicates a transition process in this region. A cyclone of depth 990 mb rises from the south, and subsequently the col should move away and the warm sector of the cyclone should move in. This occurrence actually corresponds to class A_{II}.

The radiation field for 8 June corresponds to class A_{III}, which is the class of multicell depressions. The synoptic map shows a

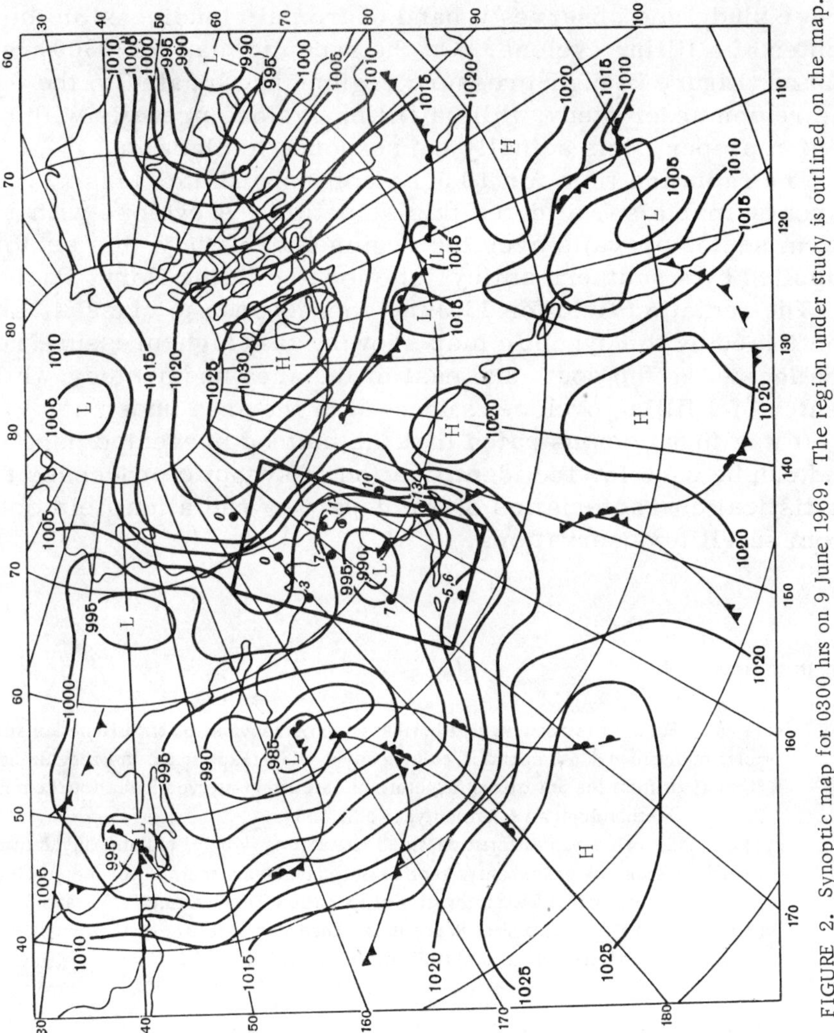

FIGURE 2. Synoptic map for 0300 hrs on 9 June 1969. The region under study is outlined on the map.

filling cyclone, which belongs to a multicell depression passing from southwest to northeast, corresponding to class A_{III}.

The radiation field for 9 June corresponds to class A_{I} in which high-pressure regions with a predominance of middle clouds and clear skies are located to the south and north (to the east and west) of the region under study. Here, almost everywhere in the region under study, one observes a band of frontal cloudiness or the center of a filling cyclone. On the synoptic map for 0300 hrs on 9 June (Figure 2) high-pressure regions are located to the east of the region under study, while a filling cyclone prevails within the region proper; this actually corresponds to class A_{I}.

The radiation field for 10 June corresponds to class A_{II}. The synoptic map also validates this selection. A cyclone with a large warm sector prevails over the region under study, this sector being supplied by a southern anticyclone located in the North Pacific.

The radiation field for 11 June corresponds to class A_{I}, and is confirmed by the synoptic map showing that high-pressure regions are located to the south and east of the area under study, while the center of a filling cyclone is located in the area under study.

It was thus demonstrated that the method presented here is valid and can be used for the identification of synoptic processes from statistical characteristics of the departing radiation field obtained from satellite observations.

BIBLIOGRAPHY

1. Tolstikov, E.E. Opyt raspoznavaniya baricheskikh obrazovanii po statisticheskoi strukture polei radiatsionnykh temperatur, poluchennykh s ISZ (Experience in Recognizing Baric Formations from the Statistical Structure of Satellite-Observed Radiation Temperature Fields).— Meteorologiya i Gidrologiya, No.3. 1969.
2. Tolstikov, E.E. Issledovanie statisticheskikh kharakteristik polei radiatsionnykh temperatur i nekotorye voprosy primeneniya teorii raspoznavaniya v sputnikovoi meteorologii (A Study of the Statistical Characteristics of Radiation Temperature Fields and Some Problems of the Application of Recognition Theory in Satellite Meteorology).— Author's abstract. Gidrometeorologicheskii Tsentr SSSR. 1970.

UDC 551.509

EVOLUTION OF FRONTAL CLOUDINESS AND CYCLONES DURING THE SUDDEN INVASION OF COLD AIR MASSES

Yu. V. Kurilova

The paper examines the evolution of an exceptionally active synoptic process over moderate latitudes of European USSR, and of processes in other latitude bands. Use is made of averaged synoptic parameters characterizing air masses, their transformation, contrasts at fronts, and supplementary satellite information — nephanalysis, and cloud top heights derived from radiation measurements. Considered are features of cloud fields and of their precipitation in cyclones at various latitudes as a function of the characteristics of air masses and frontal interfaces.

The synoptic process studied below occurred over European USSR from 13 to 18 July 1970 and was extremely active; it was accompanied by intensive precipitation. The high baric field during these days had a thick crest over Eastern Europe and troughs over Western Europe and Eastern Siberia. A similar case of a sudden cold invasion was discussed in /2/.

The evolution of the process was induced by a heavy removal of heat at the rear of this crest, first to the north and then to the northeast, in northern parts of Western Siberia. This high crest, which actively propagated to the northeast, and a weak crest over Central Asia caused the trough over Western Siberia to gradually bend eastward at its central part. This process terminated in the separation of an isolated cold center, which determined the development of a cyclone over Central Asia from 15 to 18 July.

The second large-scale trough over Western Europe was situated meridionally and, due to the western edge of the crest over Eastern Europe, it did not move easterly past 20°E. As of 15 June a series of cold invasions invaded the rear of this meridional trough; this is confirmed by the appearance on relative topography maps of closed cold centers and their gradual movement to the south: from 52°N on 15 July, to 47°N on 16 July, to 44°N on 18 July (Figure 1). In connection with this the principal contrasts on the relative topography maps also moved southerly and increased by up to 12 decameters over 1000 km on 15 July and 20—24 decameters over 1000 km on 16—17 July (Table 1).

TABLE 1. Averaged atmospheric parameters for regions of frontal zones and air

Date	Type of front	T °C								Relative topography of 1000/500 mb			K° 1000 km			
		cold air				warm air										
		ground	850	700	500	ground	850	700	500	H, decameters	φ °N	Γ decameters/10³ km	ground	850	700	500
													Secondary			
13	SCF	16	4	—5	—21	22	—8	—3	—17	548	57		4	4	3	4
14	SCF	21	8	—5	—16	24	11	0	—16	558	50		3	3	3	0
													Cyclone over			
13	CF	27	10	—3	—16	30	16	2	—14			8	3	5	0	2
14	CF	28	10	—1	—14	32	18	2	—14	560	51	12	4	6	3	0
15	CF	30	12	1	—13	35	19	5	—12	560	47	16	5	5	4	1
17	CF	32	18	4	—12	31	23	8	—11	568	44	12	1	5	4	0
18	CF	24	19	6	—12	21	22	11	—8	564	42	8	0	3	5	4
	CF															
													Cyclone occlusion fronts of			
14	OF	18	9	—1	—17	20	7	—2	—2				2	2	1	0
15	OF	17	6	—4	—21	18	7	—3	—19				1	1	1	2
16	OF	10	3	—5	—20	13	8	—2	—16				3	5	3	4
													Northern latitude			
15	WF	16	6	—2	—16	26	13	2	—12	540	65	12	10	7	4	4
15	CF	15	6	0	—13	26	14	2	—12	540	50	16	9	6	2	5
16	WF	13	7	—2	—18	23	24	3	—18				10	7	6	3
16	CF	15	6	—3	—15	23	14	3	—13	548	57	12	8	7	5	2
17	OF	16	5	—4	—19	20	11	2	—12			12	4	6	6	5
18	OF	18	6	—3	—20	16	4	3	—12			12	2	6	6	8
													Waves above			
15		3	5	—1	—15	26	6	3	—21	544	46	20	8	11	4	3
16		13	6	—2	—15	20	13	3	—12	544	50	24	7	7	5	5
17		14	6	—2	—15	30	16	4	—11	540	43	20	16	10	6	4
18		14	7	—3	—15	25	17	3	—11	548	50	16	11	10	6	4
													Stationary			
13	CF	10	0	—4	—17	12	6	—3	—18				12	2	6	1	1
14	CF	10	2	—3	—17	15	8	—3	—17				8	5	6	2	0
15	WF	11	—1	—6	—5	17	4	—4	—4				8	6	5	2	1
16	WF	8	—3	—5	—21	14	5	—3	—16				12	6	8	2	5
17	WF	10	2	—5	—10	16	7	—1	—10				8	4	4	3	3
18	CF	16	1	—4	—19	17	4	—4	—19				6	1	3	0	0

Note. SCF denotes secondary cold front, CF cold front, OF occlusion front, and

masses during July

	$d\,°\mathrm{C}$ cold air				$d\,°\mathrm{C}$ warm air				H_t km	τ 10 mb/12 hrs			Precipitation			tan α
	ground	850	700	500	ground	850	700	500		850	700	500	N	R, mm	$S_{R>10}$	
fronts																
	5	3	3	3	12	5	5	9	6.5	1	2	2	11	16		0.010
	10	4	4	5	12	7	5	11	6.5	0	1	1	5	7		0.005
Central Asia																
	19	8	4	10	16	2	7	3	5.5	1	3	3				0.005
	17	7	9	11	21	16	6	6	7.0	0	0	1	2			0.005
	12	7	5	10	18	20	7	2	7.0	1	2	3	3			0.003
	22	14	10	6	21	16	9	13	8.0	0	1	2				0.008
	10	12	6	8	6	14	10	15	5.0	1	1	2	3			0.008
northern latitudes																
	6	7	3	7	9	4	6	6	5.0	0	2	2	1			
	6	4	2	5	5	3	2	6	5.0	1	3	3				
	1	5	2	6	1	2	6	5	5.0	1	2	3				
cyclone																
	3	3	2	4	10	4	6	10	5.5	5	5	6	6	32	200	0.010
	6	3	2	7	10	4	6	10	8.0	5	9	10	9			0.007
	3	3	1	4	10	5	7	9	7.5	3	6	7	7	56	100	0.000
	5	6	8	6	10	5	7	9	8.0	2	8	10	3			0.007
	10	2	3	8	15	5	3	8	6.5	2	6	7	4	5		0.010
	5	4	6	8	2	6	9	10	4.5	1	0	0		25		0.007
cold front																
	0	3	4	7	6	6	4	7	9.5	3	6	7	18	25	300	0.009
	2	3	4	5	6	6	3	4	9.0	4	5	6	16	36	300	0.010
	5	2	2	3	5	4	7	7	8.0	4	8	10	23	51	400	0.010
	3	3	1	4	10	9	4	14	10.0	4	8	9	16	51	350	0.005
front																
	2	15	1	6	3	6	5	4	4.5	1	2	2	3	1		0.005
	5	5	5	8	4	5	5	8	7.5	1	2	3	2	2		0.007
	1	7	12	10	1	5	6	8	7.5	2	3	6	3	6		0.003
	2	1	0	3	3	2	1	4	6.0	2	3	4	5	6		0.003
	1	3	2	7	10	3	3	7	6.0	−2	−2	−3	7	10		0.006
	5	3	7	9	5	3	5	11	5.0	−1	3	−5		2		0.003

WF denotes warm front.

This sudden sharpening of thermal contrasts in the lower half of the troposphere was responsible for the type of development of atmospheric processes over European USSR with intensive precipitation over wide areas. After sharpening of contrasts on 15 July, associated with the first cold invasion, a cyclone arose in northern latitudes of Europe, and after sharpening on 16 July over southern European USSR an active wave appeared at its cold front. The entire synoptic process under study thus reduced to the evolution of four synoptic formations: cyclone moving rapidly in northern latitudes; active waves developing at the cold front in moderate latitudes; stationary front of northern latitudes; and a cyclone over Central Asia.

An analysis of the evolution of cloudiness associated with each of the above synoptic formations is studied below, as well as the nature of the relationship between precipitation and this cloudiness.

Figure 1 shows schematically the synoptic processes: 500-mb maps show the locations of fronts, zones of continuous cloudiness according to nephanalysis, radiation minima limited by the -20° isotherm, zones of precipitation, as well as cold centers from relative topography maps. Table 1 lists the following average quantities, characterizing air masses and fronts: temperature T and dew-point deficit d of cold and warm air masses, depth H and latitude φ of the cold center on relative topography maps, cloud top height H_t, vertical motions τ, number N of precipitation-recording stations, intensity R of the latter and area with precipitation in excess of $10\,\mathrm{mm}$ ($S\,\mathrm{km}^2$), as well as temperature contrasts K in the frontal zone at different levels. All the quantities characterizing air masses and fronts were calculated from 10—30 values.

At the start of the meridional cold-air invasion (13 July) several isolated and shallow cold centers formed at 50—55°N and produced secondary fronts. On 14 July the cold center moved eastward and combined with a weak cyclonic center over Central Asia, activating the latter. This manifested itself in substantial expansion of the cyclone's cloudiness zone (Figure 1).

Secondary fronts, moderately high with a small angle of inclination (tan α equal to 0.005 and 0.010) and weak temperature contrasts at these levels not exceeding 4°, were accompanied on 13 July by an extensive cumulus-covered region and precipitation at 11 stations. However, the intensity of this precipitation was small, with maxima of 16 mm attained only at individual stations. The precipitation was induced by unstable stratification behind the cold front during the invasion of a humid ocean air mass and its rapid motion over the heated continent. The dew-point deficit in the cold mass on 13 July was small (3—5°).

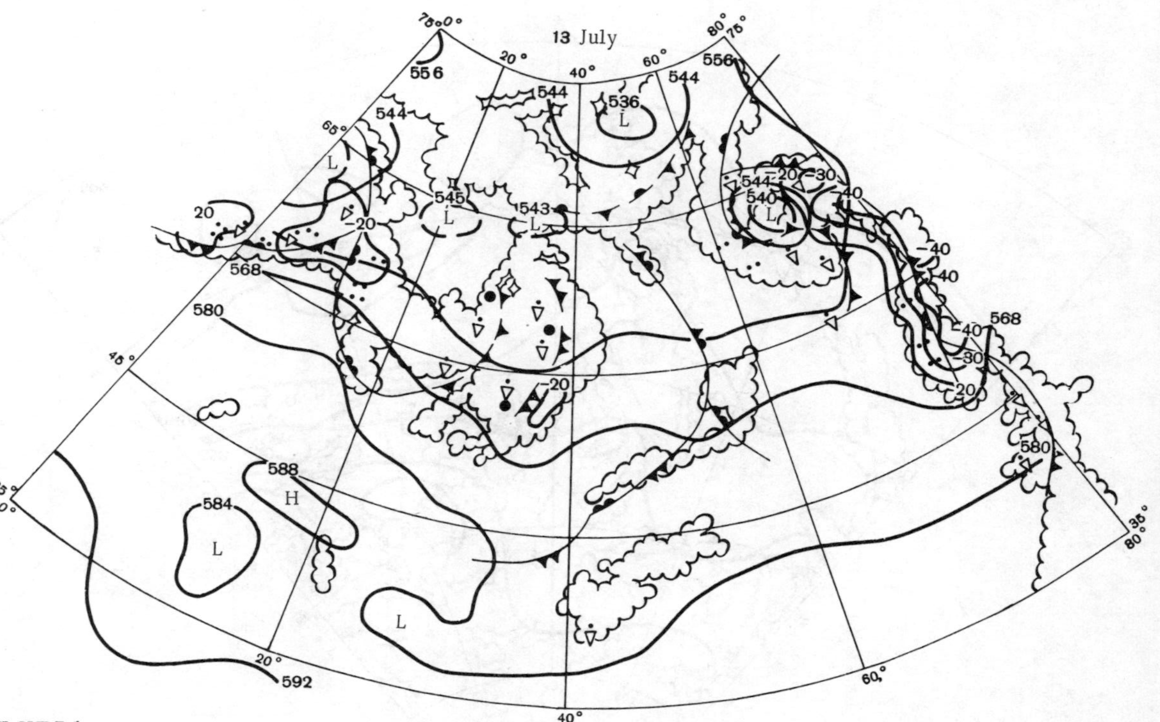

Figures 1a — f illustrate the evolution of baric fields of cloud and frontal systems from 13 to 18 July 1970.   Black full circles denote cold centers taken from the relative topography map.

14 July
580
568
556
544
560
L
532
L
540
544
556
568
580
592
584
H
584
H
L
580
582
L
H
584
580
−20
−20
−20
−20
80°
45°
60°
75°
90°
65°
80°
80°
60°
40°
20°
0°
40°
20°
0°
30°
35°
75°
85°
85°

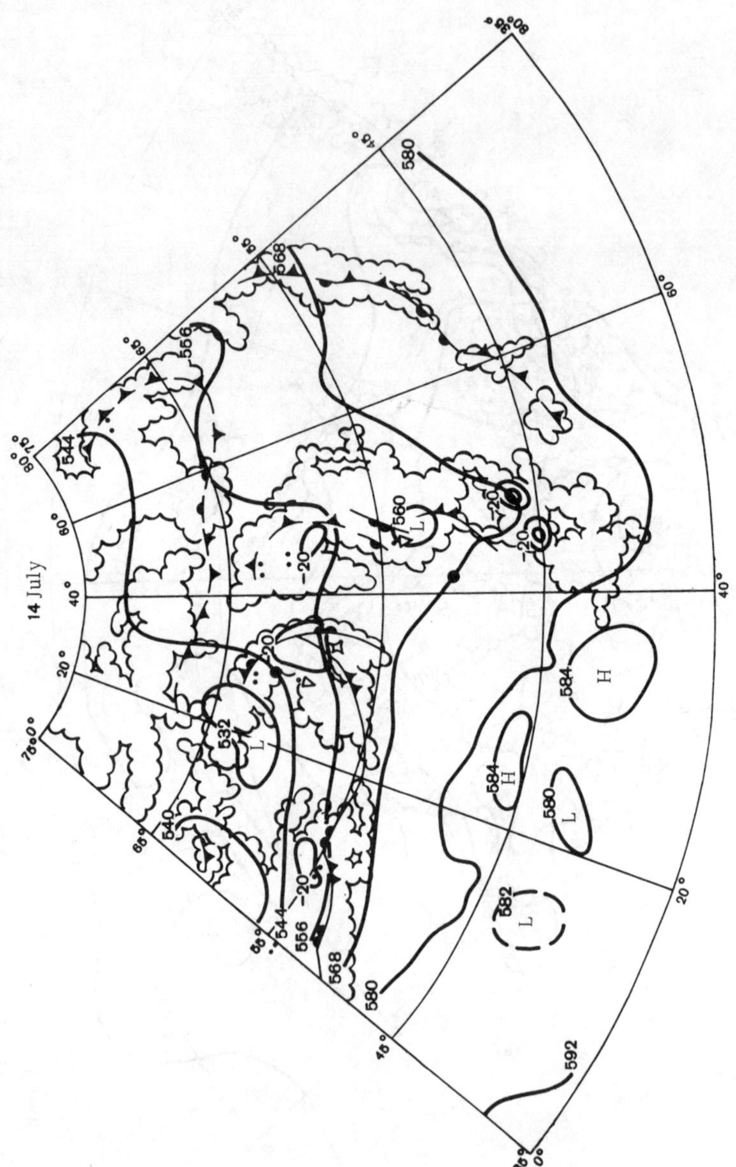

FIGURE 1b.

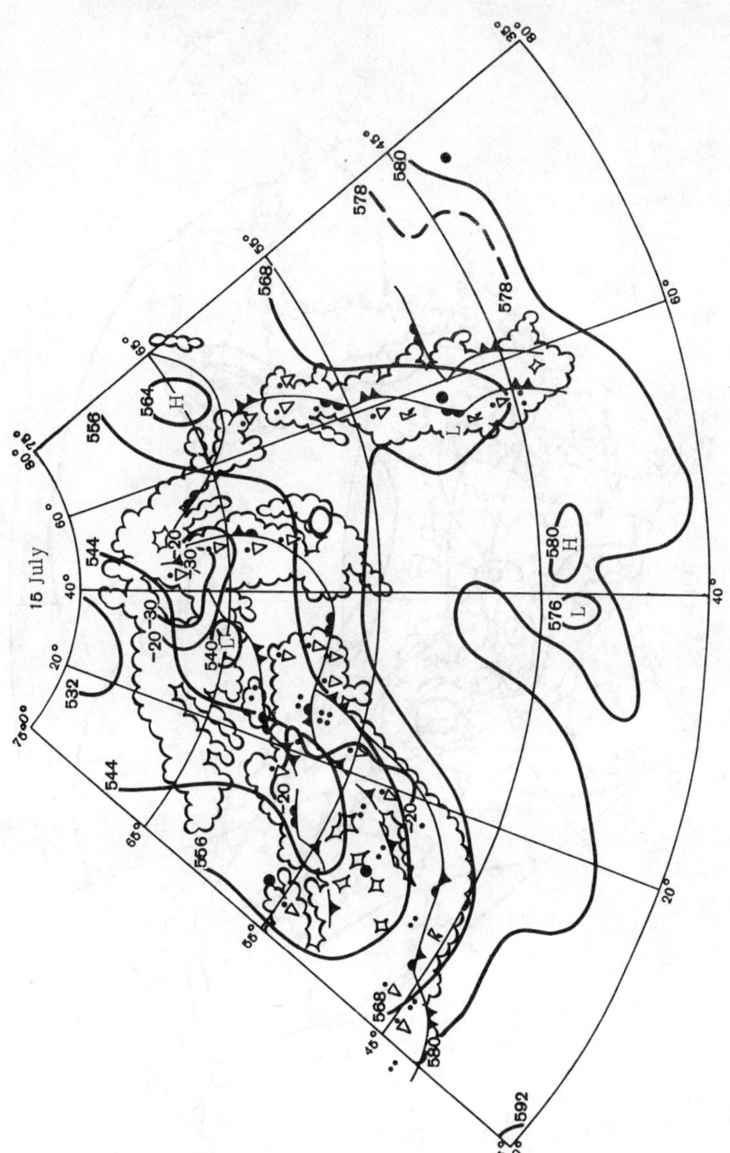

FIGURE 1c.

16 July
580
580
568
564
556
544
532
556
544
540
548
552
552
568
580
592
H
L
L
L
L
L
L
80°
60°
40°
20°
45°
36°
80°
75°
70°
65°
60°
55°
50°
40°
20°

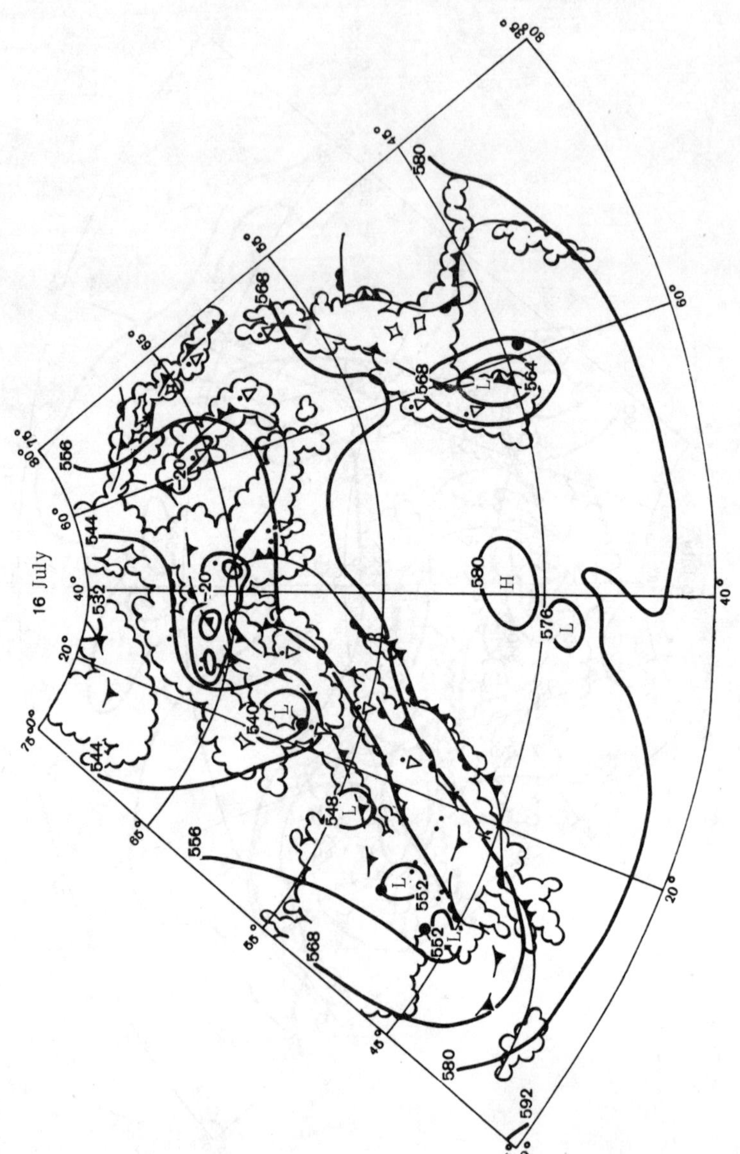

FIGURE 1d.

17 July
580
576 H
564 L
568
560 L
556
540 L
544
552 L
552 L
556
560 H
568
580
583 H
580
580 H
580 H
580 H
580 H
L
L
-20
-30
-40
40 40
-30
-20

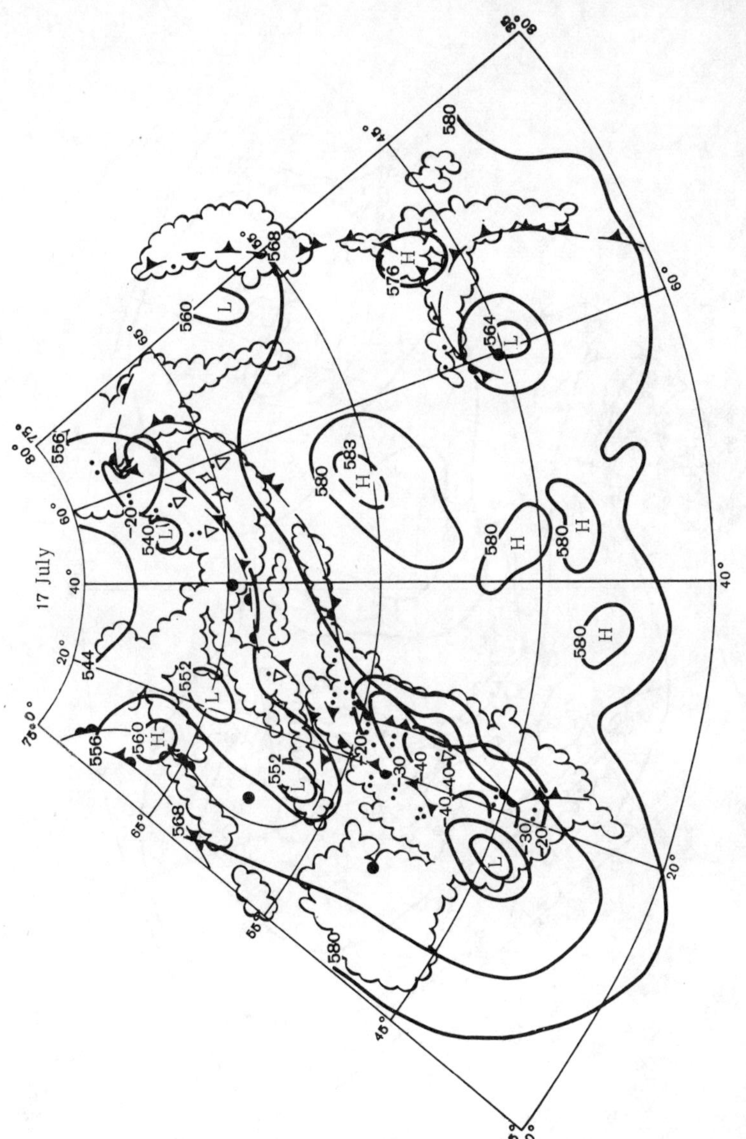

FIGURE 1e.

18 July
548
556
568
556
568
584
H
564
L
556
L
568
580
556
568
580
-20
-30
-40
-40
-30
-20
L
-40
R
R
580
568
L

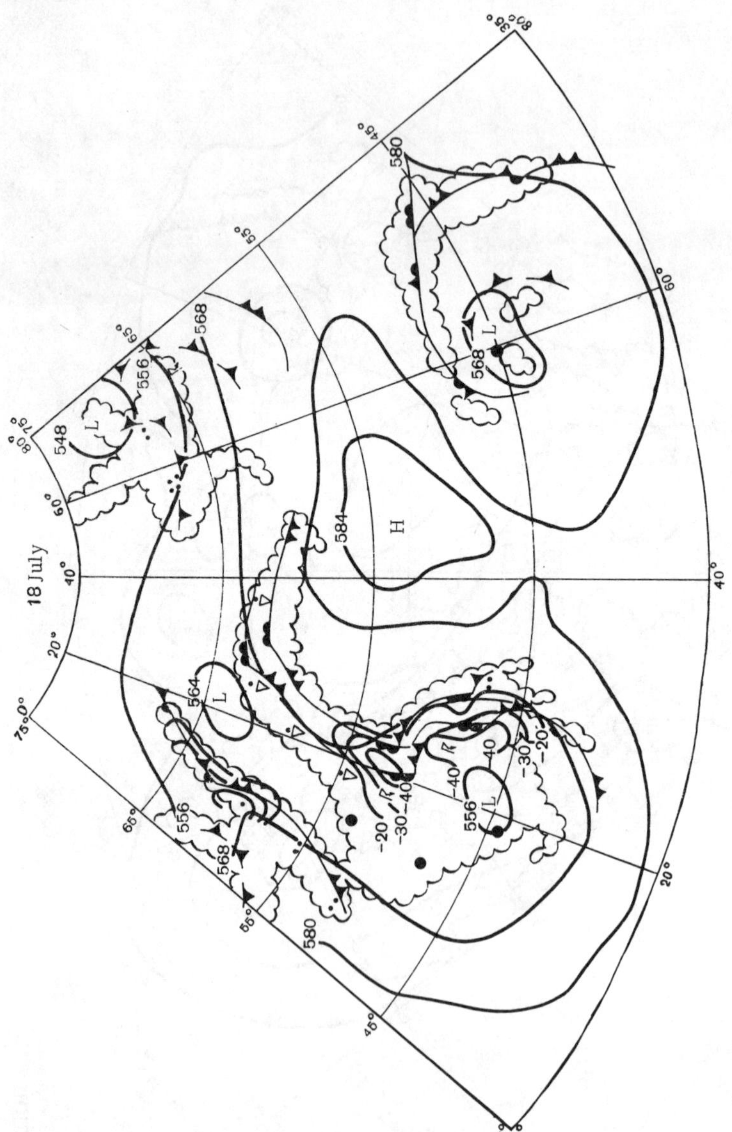

FIGURE 1f.

As it moved into the continent, the cold sea air was heated during 13 to 14 July from 16 to 21° at the surface and from -21 to -16° at the 500 mb level; the dew-point deficit at its surface increased perceptibly (from 5 to 10°). Correspondingly, the area occupied by cloudiness and thunderstorms decreased markedly (from 11 to 5 stations), and the precipitation intensity also decreased (from 16 to 7 mm during 12 hours).

The cloud top heights obtained from satellite radiation measurements in the 8–12 μ range, subsequently termed the radiation height of clouds, remained (owing to some inertia of the latter) identical and equal to 6.5 km.

The cold region, which induced a secondary front on 13 July, moved further eastward and on 14 July strongly activated a cyclone being damped over Central Asia. Regeneration of this cyclone is confirmed by substantial widening of the cloud band (Figure 1) and by the moderate increase in its thickness (Table 1). This wide meridional cloud band started narrowing and as if "melted" only on 16 July, when the regenerated cyclone started occluding.

Transformation of the cold air mass moving over the continent to the southeast from 13 to 17 July was found to be quite substantial, and the temperature at the ground and at the 850 and 700 mb levels rose from 16 to 32°, 4 to 19° and -5 to 6°, respectively. The cold air mass became increasingly drier during the transformation, so that from 17 July its deficit attained 22° at the surface and 10–14° in the lower troposphere. The warm air mass was even drier. The reduction in temperature contrasts and increase in dew-point deficit following 15 July coincides with the inception of the degeneration of the cloud masses.

The radiation height of the central part of the cloud mass on the first days following regeneration of the cyclone increased from 5.5 to 7 km, and then over a number of days remained substantial, irrespective of the reduction in temperature contrasts and increase in dew-point deficit. The latter points to a marked inertia of cloud masses.

The high air dryness, typical of Central Asia, is also responsible for the low precipitation rate, even when the cloud has a marked inclination and the cloudiness is high.

The nature of air masses, particularly their transformation, is shown by this example to play an important role in estimating cloudiness and its precipitation.

The principal synoptic process over European USSR is the rapid motion to the north of the continent of a series of rapidly occluding cyclones. The first of these is seen in Figure 1 as an occluded cyclone with very small contrasts (1–4° per 1000 km) and with a dew-point deficit of 1–7°. The cloudiness fields according to nephanalysis data narrowed down perceptibly from 14 to 16 July, the radiation height remaining unchanged and equal to 5 km (Table 1).

The development of the following cyclone moving in northern latitudes is seen more clearly. Having first appeared at 10°E in the form of a wave, this formation became a young cyclone as early as 15 July and, involving a wide warm sector, reached 28°E. It was finally observed on 17 July as an occluded cyclone at 70°E.

The temperature of the cold air mass associated with this invasion rose somewhat in connection with transformation of air during 15–18 July at the surface, but decreased slightly at the 500 mb level, which points to a cold advection continuing at high altitudes. At the same time the temperature contrast in the cyclone occlusion stage decreased perceptibly, while at high altitudes it was retained and even increased somewhat. In this example the warm front had the greatest contrasts at the surface. The lowest values of the dew-point deficit (1–4°) in this young cyclone were also concentrated in the region of the warm front. These two factors are associated with a wide area of warm-front cloudiness, which is atypical. This asymmetry is particularly perceptible on 16 July (Figure 1). The cold front, at which the dew-point deficits were large (5–8°), remained owing to this predominant cloudlessness.

The cloud top height according to radiation data in the young cyclone on 15 July was 8 km, whereas at the last stage it was 4 km (Table 1).

The precipitation zone of 15 July during the young-cyclone stage was quite wide, but narrowed down with occlusion of the cyclone. The precipitation intensity also decreased substantially from the time of occlusion (17 July), from 56 to 5 mm per 12 hours (Table 1). Precipitation ceased on 18 July.

The above process in general is not too active; this is confirmed by moderate temperature and dew-point deficit contrasts. The process most active in the summer of 1970 was that occurring over southwestern European USSR, where a wave formation accompanied by intensive thundershowers (up to 51 mm per 12 hours) occurred at the cold front of the above cyclone.

The first wave at the cold front arose on 15 July, a system of several weak waves appeared on 16 July, two clearly delineated waves formed on 17 July, while on 18 July, when the cyclone at the cold front of which the waves developed had weakened, a third wave appeared at its northern part. These three waves bent the front so that it as if formed a wide warm sector, associated with intensive northward propagation of the anticyclonic crest.

The exceptional activity of the wave-formation process at the cold front was induced by subsequent penetration of cold air into the deep south (540 decameter center at 43°N) which induced (as can be seen on the relative topography map) a new increase in the contrasts to 20–24 decameters per 1000 km, and at the surface and 850 mb level to 16 and 10° per 1000 km, respectively. The sharp rise in contrasts at the front is accompanied by sharp dew-point deficits, such as 1–2° at the 700 mb level (Table 1).

The extremal values of temperature contrasts and the high saturation of the air mass were responsible for the exceptionally wide fields of intensive precipitation. Thus, precipitation with intensity of more than 10 mm per 12 hours occupied more than 400 km^2 (with a maxima of up to 51 mm per 12 hours).

The cloudiness field also evolves in accordance with the variation in temperature contrasts. On 15 July nephanalysis maps show that the cloudiness of the northerly cyclone and of the newly arisen wave forms a single extensive and wide band, which subsequently separates into two masses. Here the cloud mass in the north starts to break up, while the cloudiness of the newly arisen wave at the south starts to strengthen. Initially circularly shaped, as of 18 July this cloudiness was bent by the three waves at the cold front into a curved band, similar to the cloudiness of the cyclone before occlusion. On 17 and 18 July, a high-level (9–10 km) center adjoining the crests of the front's waves began to form.

The previously mentioned stationary front is located near the shore line of the continent. It was followed from the 850 mb level and had a slight inclination (tan α = 0.003). During the period under study it prevailed all the time and its location hardly varied. This front is expressed by several contrasts only at the ground and at the 850 mb level. The existence of this stationary front is associated primarily with temperature contrasts of various surfaces (land-sea).

According to nephanalysis data, the cloudiness at this front does not always form a continuous band but, as the front proper, it is activated at individual sections and greatly changes its configuration on individual dates. Activation of the front at individual segments is apparently induced by convective instability, associated with the movement of cold air to the continent. Radiation cloud heights at this front are small and are quite variable. The significance of these stationary fronts in the precipitation field is determined to a large extent by the humidity characteristics of the cold sea air. On days when the dew-point deficits at low levels are least (1–3°), precipitation with an intensity of 10 mm per 12 hours occurs at a large number of stations (5–7 stations).

TABLE 2. Average values and limits of characteristics of air masses and fronts of synoptic processes of the same type over European USSR in July 1970

Synoptic process	H_t	RT^{500}_{1000}	$\bar{K}$				τ			$\bar{d}$				Number of stations with precipitation	$\tan a$
			Surface	850	700	500	850	700	500	Surface	850	700	500		
Average values															
Cyclone of northern latitudes	7	13	9	7	5	4	4	8	10	4	4	3	5	6	0.009
Wave at cold front	9	20	10	9	5	4	8	7	5	3	3	3	5	18	0.010
Cyclone over Central Asia	7	10	3	4	3	1	1	0	0	16	10	7	9	2	0.006
Stationary front of northern latitudes	5	8	4	5	1	1	2	3	2	3	4	6	7	4	0.004
Occlusion fronts	7	6	2	3	3	3	2	4	5	7	4	3	6	1	
Limits of variations															
Cyclone of northern latitudes	5–8	12–16	2–10	3–7	2–6	2–8				3–6	3–6	3–7	3–8	2–7	0.007–0.010
Wave at cold front	8–10	16–20	8–10	8–10	5–6	4–5				0–4	2–3	1–4	3–7	16–23	0.005–0.010
Cyclone over Central Asia	5–8	8–16	0–5	3–6	3–5	0–4				10–20	4–10	4–10	4–11	1–5	0.002–0.005
Stationary front of northern latitudes	4–6	6–12	1–6	3–8	1–2	1–5				1–5	3–10	2–7	3–9	3–7	0.003–0.006
Occlusion fronts	5–6		1–3	1–5	1–3	0–4				2–11	3–10	2–5	5–9	3–7	0.003–0.008

The differences in the synoptic processes of various latitudes analyzed above are particularly clear from variables averaged over the limits of synoptic processes of the same type, tabulated in Table 2. Correlation of these data shows that the most active synoptic wave-formation process at cold fronts of moderate latitudes corresponds to maximum temperature contrasts ($\bar{\Gamma} = 20$ decameters per 1000 km on the relative topography map and $\bar{K} = 10°$ at the ground) and to minimum dew-point deficits ($\bar{d} = 3°$). This process is also associated with high clouds ($\bar{H_t} = 9$ km), accompanied by intensive precipitation over a large area: more than 10 mm per 12 hours were recorded at 16–23 stations, the maximum precipitation intensity reaching 51 mm/12 hours.

The low stationary front of northern latitudes near the shore had moderate temperature contrasts (4–5°) and very small average dew-point deficits at low levels (about 3°). This front corresponds to quite large, although discontinuous fields of middle clouds with radiation heights of about 5 km, which produced thundershowers with a maximum intensity of 10 mm/12 hours, but only at individual points.

The cloudiness in the cyclone over Central Asia at moderate, but somewhat smaller contrasts (3–4°), and with markedly higher dew-point deficits (10–16°), was sufficiently high (7 km). Vertical motions, extremal for the active process of wave formation in moderate latitudes, are close to zero for the process over Central Asia (Table 2). The quite high cloudiness in the cyclone over Central Asia at small temperature contrasts, high deficits and zero vertical flows can be attributed to the fact that it is breaking up. This is confirmed by the narrow, diffuse, and as if "melting" cloud band on the nephanalysis map (Figure 1) and slight precipitation, observed only at individual stations.

Analysis of the evolution of this process over European USSR points to the existence of certain dependences of the temperature contrasts, vertical flows, dew-point deficits and number of stations with precipitation on the radiation cloud height, i. e., the numerical characteristic obtained from satellites. These relationships were also examined by means of correlation graphs, where quantities averaged for the synoptic formation were taken as one of the values. Due to the small number of values used, we did not calculate the correlation coefficients.

The radiation height of cloudiness correlates positively with the vertical flows, although the scatter of values is quite substantial. The correlation becomes closer when the values of the vertical flows are taken at one fixed level. The correlation is weakened perceptibly by a group of points with substantial radiation heights and small vertical flows, corresponding to the break-up of cloudiness of occlusion fronts and fronts of a cyclone situated over Central Asia in which the cloud formation has already ceased. The latter circumstances make it difficult to use the radiation height of clouds as a parameter in estimating vertical flows.

The contrasts of temperatures at the front and the radiation cloud top heights also correlate, since they are related to vertical flows: the former as a cause and the latter as an effect. A closer correlation exists between the radiation cloud top height and temperature contrasts at the surface and at the 850 level. This relationship vanishes at higher levels.

The radiation cloud top height gives some idea about the thickness of clouds and, as shown in the above analysis, an approximate idea concerning temperature contrasts in the frontal zone and vertical flows. It was shown, however, that the radiation cloud top height does not aid in judging about precipitation.

The only way of singling out those groups of synoptic processes with which quite thick clouds, accompanied by heavy precipitation, are associated is by using a combination of the radiation cloud top height and the dew-point deficit. Comparison of the ranges of variation in temperature contrasts, the dew-point deficit and the radiation cloud top height with the ranges of variation in precipitation for the synoptic processes under study enables one to obtain critical parameters for identifying active synoptic processes accompanied by heavy precipitation. The most active synoptic process of wave formation in moderate latitudes corresponds to a dew-point deficit ranging from 0 to 4° at all levels below 500 mb with a very high radiation cloud top height (8—10 km). In this case precipitation extends over a large area, so that from 16 to 23 stations recorded precipitation more than 10 mm/12 hours.

The less active northern cyclones have corresponding to them a wider range of dew-point deficits (3—8°) and smaller radiation cloud top heights (5—8 km). During the most active stage of development of this cyclone the extremal values of precipitation intensity for two days were as high as 32 and even 56 mm for 12 hours, but areas with precipitation in excess of 10 mm were in this case relatively small. The precipitation abated on subsequent days and intensities in excess of 10 mm/12 hours were observed only at individual stations.

The radiation cloud top height for the stationary front of northern latitudes with its spatially and temporally varying broken cloudiness is more conservative and equal to 4—6 km with a variation of 0—10° in the dew-point deficit. Precipitation in this process was slight; it was recorded at only 3—7 stations and as a rule was less than 10 mm.

Far greater dew-point differences were observed in the cyclone over Central Asia, where the cloud top height was quite high (from 5 to 8 km), temperature contrasts were relatively small, and vertical flows were close to zero. There was virtually no precipitation in this cyclone.

It is thus seen that an additional variable estimating the hygrometric properties would, along with the radiation cloud top height, have made it possible to estimate the activity of the process from the standpoint of precipitation.

Variations in the humidity characteristics over the microwave range can be used to estimate intensive synoptic processes only with the aid of satellite data.

TABLE 3. Average surface (T) and radiation (T_r) temperatures, differences (Δ) between them, and the dew-point deficit (d) for cloudless conditions in the crest during July

Date and hour	T	n	T_r	n	Δ	d	n	Coordinates $\varphi°$	$\lambda°$	Synoptic situation
13 July 1500 hrs	20	11	12	20	8	10	11	48	50	Zonal flow
15 July 1500 hrs	29	10	26	22	3	22	11	48	45	Foward part of crest
16 July	15	8	11	51	4	5	5	53	45	Rear of crest
0300 hrs	18	11	12	40	6	8	12	50	45	Crest axis
17 July	24	5	13	30	11	11	5	58	45	Rear of crest
0300 hrs	23	10	12	38	11	9	10	58	55	Crest axis
17 July	31	15	24	36	7	20	13	52	55	Forward part of crest
1500 hrs	30	10	17	30	13	13	10	58	55	Rear of crest
	32	14	25	40	7	21	16	53	45	Crest axis
	35	8	31	13	4	30	6	46	42	" "
18 July	22	6	12	68	10	3	6	57	70	Rear of crest
0300 hrs	21	11	13	36	12	7	11	57	45	" " "
	26	11	15	53	11	12	12	52	45	" " "
	23	5	14	14	9	8	5	55	55	" " "
20 July 1500 hrs	32	9	27	34	5	22	8	53	50	Forward part of crest
21 July 1500 hrs	23	5	14	30	9	11	5	50	40	Rear of crest
Average	32		28		4	25				Forward part of crest
values	19		12		7	7				Crest axis
for processes of the same type	25		15		10	11				Rear of crest

Under the conditions of the cloudless anticyclone, which was situated on 16 July over the east of European USSR and up to 21 July moved further northward, the radiation temperature was used for estimating the air temperature at the surface. For this purpose averaged radiation temperatures, averaged air temperatures at the surface and averaged dew-point deficits were determined for cloudless sections from nephanalysis maps. Table 3 shows that the forward part of the crest has corresponding to it high dew-point deficits and temperatures. The average surface temperature here is 32° and the average radiation temperature is 28°. The average values of the dew-point deficit (25°) point to high air dryness. Correspondingly, the radiation temperatures in the forward part of the crest are close to those at the surface, and the difference between them is only 4°.

In the rear part of the crest, adjoining its axis, the temperatures are much lower, the surface temperatures are 25°, while the radiation temperatures are only 15°. The air here is more humid and the average difference between the surface and radiation temperatures is 10°. The differences in the humidity of air masses of the forward and rear parts of the crest (as can be seen from this example) determine the magnitude of corrections to the radiation temperature in the rear and forward parts of this crest.

This difference in moisture content at different peripheries of the crest is also confirmed by the distribution of cloudiness relative to the axes of high troughs and crests, obtained in /1/, where it was shown that cloud systems are usually concentrated in the rear part of the crest while, as a rule, its forward part is cloudless. This situation is demonstrated in Figure 1.

Our analysis thus shows that the degree of atmospheric humidity can be estimated by comparing ground and radiation temperatures under cloudless conditions.

BIBLIOGRAPHY

1. Kurilova, Yu.V. and I.R. Egorova. Nekotorye osobennosti evolyutsii tsiklonov i ikh oblachnykh sistem nad Tikhim okeanom (Some Features in the Evolution of Cyclones and their Cloud Systems Over the Pacific).— Trudy Gidrometeorologicheskogo Tsentra SSSR, No.73. 1971.
2. Minina, L.S. and I.A.Maklakov. Evolyutsiya tropopauzy pri rezkom vtorzhenii kholodnoi vozdushnoi massy (Evolution of the Tropopause During a Sudden Invasion of a Cold Air Mass).— Trudy TsIP, No.152. 1966.

EXPLANATORY LIST OF ABBREVIATIONS APPEARING IN THIS BOOK

Abbreviation	Full name (transliterated)	Translation
AN SSSR	Akademiya Nauk SSSR	Academy of Sciences of the USSR
DAN SSSR	Doklady Akademii Nauk SSSR	Reports of the Academy of Sciences of the USSR
GGO	Glavnaya Geofizicheskaya Observatoriya	Main Geophysical Observatory
GMTs	Gidrometeorologicheskii Nauchno-Issledovatel'skii Tsentr	Hydrometeorological Research Center
GOIN	Gosudarstvennyi Okeanograficheskii Institut	State Oceanographic Institute
MMTs	Mezhdunarodnyi Meteorologicheskii Tsentr	World Meteorological Center
NIIAK	Nauchno-Issledovatel'skii Institut Aeroklimatologii	Scientific Research Institute of Aeroclimatology
TsAO	Tsentral'naya Aerologicheskaya Observatoriya	Central Aerological Observatory
TsIP	Tsentral'nyi Institut Prognozov	Central Weather Forecasting Institute
VNMS	Vsesoyuznoe Nauchno-Meteorologicheskoe Soveshchanie	All-Union Scientific Meteorological Conference